Popular science and public opinion in eighteenth-century France

Manchester University Press

STUDIES IN EARLY MODERN EUROPEAN HISTORY

This series aims to publish challenging and innovative research in all areas of early modern continental history. The editors are committed to encouraging work that engages with current historiographical debates, adopts an interdisciplinary approach, or makes an original contribution to our understanding of the period.

Already published in the series

The rise of Richelieu Joseph Bergin

Sodomy in early modern Europe
ed. Tom Betteridge

The Malleus Maleficarum *and the construction of witchcraft*
Hans Peter Broedel

Latin books and the Eastern Orthodox clerical elite in Kiev, 1632–1780
Liudmila V. Charopova

Fathers, pastors and kings: visions of episcopacy in seventeenth-century France
Alison Forrestal

Fear in early modern society
eds William Naphy and Penny Roberts

Religion and superstitition in Reformation Europe
eds Helen Parish and William G. Naphy

Religious choice in the Dutch Republic: the reformation of Arnoldus Buchelus (1565–1641)
Judith Pollman

Witchcraft narratives in Germany: Rothenburg, 1561–1652
Alison Rowlands

Authority and society in Nantes during the French Wars of Religion, 1559–98
Elizabeth C. Tingle

The great favourite: the Duke of Lerma and the court and government of Philip III of Spain, 1598–1621
Patrick Williams

Popular science and public opinion in eighteenth-century France

MICHAEL R. LYNN

Manchester University Press
Manchester and New York

distributed exclusively in the USA by Palgrave

Published by Manchester University Press
Oxford Road, Manchester M13 9NR, UK
and Room 400, 175 Fifth Avenue, New York, NY10010, USA
www.manchesteruniversitypress.co.uk

Distributed exclusively in the USA by
Palgrave, 175 Fifth Avenue, New York,
NY10010, USA

Distributed exclusively in Canada by
UBC Press, University of British Columbia, 2029 West Mall,
Vancouver, BC, Canada V6T 1Z2

British Library Cataloguing-in-Publication Data
A catalogue record is available from the British Library

Library of Congress Cataloging-in-Publication Data applied for

ISBN 0 7190 7373 1 *hardback*
EAN 978 0 7190 7373 1

First published 2006

15 14 13 12 11 10 09 08 07 06 10 9 8 7 6 5 4 3 2 1

Typeset in Monotype Perpetua with Albertus
by Northern Phototypesetting Co Ltd, Bolton
Printed in Great Britain
by Biddles Ltd, King's Lynn

For Judy

Contents

Acknowledgements

It gives me great pleasure to acknowledge the help I received while writing this book. The opportunity for scholarly interactions at conferences, in the archives, and in cafés remains one of the greatest benefits of working within the modern-day Republic of Letters. This book started life as a dissertation at the University of Wisconsin-Madison under the aegis of Domenico Sella who patiently and graciously guided me through the rigors and joys of graduate school. Suzanne Desan and Tom Broman took the time to read many drafts of my work in progress and offered enormously valuable advice. Other faculty in both the History and the History of Science Departments – including Laird Boswell, Bob Kingdon, David Lindberg, Michael Shank, and the late Ed Gargan – gave me the benefit of their time and understanding of history. I also had excellent models to follow as an undergraduate at Pacific Lutheran University where my professors, especially Chris Browning and Phil Nordquist, exemplified the roles of scholar, teacher, and mentor.

Other friends and colleagues from Madison helped make my graduate-school experience eminently worthwhile. My thanks go out to Franca Barricelli, Kathleen Comerford, Susan Dinan, Ted Ingham, Chip Landrum, Jody LePage, Lexi Lord, Dan Margolies, Craig McConnell, David Reid, Louise Robbins, Joe Tarantowski, and Deirdre Weaver. I am enormously thankful for having known and been friends with the late David Picard. I also wish to express my gratitude to Linda Essig, Ron Larson, and Eric Nordholm.

I have discussed my work with a number of other scholars who generously spent time with me. Bernadette Bensaude-Vincent, Paola Bertucci, Christine Blondel, Greg Brown, Colin Jones, Morag Martin, Jeffrey Ravel, and Kathleen Wellman have all shared their insights into the study of early modern France and the history of science. Roger Hahn and Tom Hankins both read portions of my work in progress and gave me excellent feedback. Dena Goodman has been particularly generous with advice and support over the years. J.B. Shank provided endless encouragement – coupled with innumerable pints of beer and cups of coffee – and has continuously pushed me to argue more deeply. Gilles Chabaud, Marie Thébaud-Sorger, Oliver Hochadel, and J.B. Shank shared their work in progress with me, much to my enlightenment and edification. My colleagues at Agnes Scott College – especially Mary Cain, Penny Campbell, Julia De Pree, Violet Johnson, Kathy Kennedy, Tracey Laird, and Peggy Thompson – have also been extremely supportive and have willingly suffered through my efforts to develop this project. Tracey Laird offered advice and assistance along the way, read the entire manuscript, and, together with Brandon Laird, provided an extremely congenial environment for both work and play.

My time conducting research was made much more enjoyable thanks to the efforts of a vast array of knowledgeable archivists. In Madison, John Neu, Robin Rider, and the staff in the Rare Book Room proved invariably helpful. In France, I received considerable assistance from the staffs at the Archives Nationales, Bibliothèque Nationale (Richelieu and Tolbiac locations), Bibliothèque historique de la ville de Paris, the Archives de l'Académie des Sciences, the Bibliothèque de l'Institut, the Centre de documentation at the Musée de l'air et de l'espace, the Arsenal, the Conservatoire nationale des arts et métiers, and the Bibliothèque Mazarine. I also did research in the pleasant environments of the Bakken Library in Minneapolis, thanks to Liz Ihrig, and at the Newberry Library in Chicago.

Portions of this project were presented or discussed in a variety of venues including the Sixth International Summer School in the History of Science in Uppsala, Sweden where Tore Frängsmyr epitomized enlightened hospitality. A seminar at the Folger Shakespeare Library, led by Karen Newman, brought together scholars working on early modern Paris. I am thankful for an invitation to present a portion of my work at Emory University's Vann Seminar in Premodern History. Bernadette Bensaude-Vincent and Christine Blondel kindly invited me to participate in a workshop at the Centre de recherche en histoire des sciences et techniques at La Villette. Chapters 4 and 5 appeared in earlier versions in *Eighteenth-Century Studies* and *Isis* respectively.

I am thankful for financial support including a Bernadotte E. Schmitt Grant from the American Historical Association, a French Historical Studies/Western Society for French History Research Award, and a Robert R. Palmer Research Travel Award. The Professional Development Committee at Agnes Scott College generously gave me several summer research stipends in support of this book.

Moral support came from John and Nancy Oblanas; Nancy also translated some key passages from a Swedish travel diary for me. I would like to thank my sister Kate for her help and encouragement over the years. My parents, John and Sally Lynn, backed my decision to go to graduate school and always proved willing to help along the way. Finally, I would like to thank Judy for hanging in there with me through thick and thin, from Madison to Paris to Atlanta. Judy offered support throughout the process of research and writing, listened while I struggled to articulate arguments, and stood by me during my mad quest to publish the results. I dedicate this book to her.

1
Introduction

In his general overview to the *Tableau de Paris*, Louis-Sébastien Mercier suggested that no matter where you went in Paris, "Everywhere science calls out to you and says, 'Look.'"[1] In the eighteenth century, science entered the public imagination and became a part of the popular mentalité where it played an important mediating role in the appropriation of the Enlightenment within urban culture. Voltaire had noted this as early as 1735 when he claimed, "everyone is pretending to be a geometer and a physicist."[2] During the age of the Enlightenment, science spread from private laboratories and royal academies to the public realm where it became an integral part of urban culture. No longer the purview of elite savants or wealthy amateurs, science became an object of popularization and commodification as an increasingly large group of people began to see it as something worth spending time and money to acquire. People avidly consumed knowledge so they could sprinkle their conversations with scientific metaphors and references or, in some cases, actually utilize and implement this newfound knowledge in their lives.

Over the course of eighteenth-century France a broad array of individuals, ranging from elite savants to more dubiously inspired charlatans, presented scientific discoveries, inventions, and theories to an eager and growing audience. Innumerable opportunities existed for people to learn about different types of science. Some of the best-known and most oft-mimicked scientific displays utilized the power of electricity. In April of 1746, for example, Jean-Antoine Nollet, a professor of experimental physics and member of the Académie Royale des Sciences, transmitted an electrical shock through 180 of Louis XV's royal guards, all holding hands, while the monarch and his entourage looked on. Later, and much to the delight of the king, he simultaneously shocked 200 Carthusian monks, "volunteers" from a monastery near Paris. Even more startling, some savants debated whether or not they could electrify eunuchs. Three of the king's musicians, all castrati, underwent a series of tests; to everybody's satisfaction, they jumped as much as the other test subjects.[3] While certainly amusing, at least to the observers, these experiments also tested the power and nature of electricity while simultaneously taking advantage of the elite audience to act as verifiers for the production of scientific knowledge.

The spectrum of popular science, however, ran a wide gamut from Nollet's demonstration before an elite, royal audience to a much more philosophically suspect scientific show found in the entertainment sector near the Boulevard du Temple. Perrin, a self-styled "professor of amusing physics," performed lectures on natural philosophy to wide acclaim, even though his experiments were dominated by theatricality and spectacle. His fame rested largely on the strengths of his dog, a spaniel that could "read, calculate, solve problems, and do other physics tricks and games." Perrin's "little savant dog" occupied a central place in his physics show; women in particular applauded this multitalented animal that by 1787 could read both English and French, as well as ably assisting demonstrations of experimental physics. Perrin managed to raise himself from relative obscurity, aided by little more than a deft hand and a very talented dog, to become a mainstay in the world of scientific popularization, and he performed his experiments in the boulevard theaters with spectacular success.[4]

The audience watching Perrin, however, differed somewhat from the one attending Nollet's experiments. While both Nollet and Perrin introduced a theatrical element into their demonstrations, the intended result of their work exhibited a number of distinct differences. Nollet engaged in a certain amount of entertainment, but he also worked on knowledge production, tried to secure patronage, and attempted to disseminate new scientific discoveries. Perrin, on the other hand, appeared more deeply concerned with the commercial nature of his work; nonetheless, he, too, attempted to educate people even as he tried to nurture an audience base that would ultimately secure his livelihood. Perrin offered some scientific information to his audience although his style of presentation may have caused people to remember the performance and not the lesson. In both cases, however, the general public purchased access to science and participated, however loosely, in the dissemination of enlightened ideas in Paris.

Science as a form of cultural capital played a key role during the eighteenth century. But how did people gain access to science? And what did they do with their newfound knowledge once they had acquired it? This book explores the appropriation of science in French society and the development of an urban scientific culture. Science underwent a process of commodification and popularization during the eighteenth century as more and more individuals sought to acquire some knowledge of scientific activities and as more and more people entered public debates on science. Followers of the Enlightenment, in all of its various forms but especially in the encyclopedic trend towards making knowledge available to everybody, suggested that members of the Republic of Letters should want to apply scientific principles to their lives so that they might make rational decisions. Thus, people adopted the language and the methodology of science in order to participate in the expanding enlightened world.

The growing importance of science fueled a demand for scientific knowledge met by a group of middling savants who launched a broad array of popularizing efforts. These popularizers helped create a public interest in science while at the same time responding to the varying concerns of the newly created audience. These middling savants functioned as cultural intermediaries between their elite colleagues and their popular audience. Subsequently, once people acquired a certain level of scientific knowledge, they began to find ways to use it. As a result, individuals founded scientifically oriented clubs, participated in the development of new discoveries, and involved themselves in debates. Those people who had appropriated some science felt empowered to engage in discussions on scientific topics and help in the formation of public opinion.

A growing web of interconnections existed between Enlightenment, science, and commerce in Parisian urban culture. As Jean Ferapie Dufieu, the author of a manual for physics, claimed in 1758, "the taste for physics is so spread out in the world today that it seems necessary, in order to appear with honor, to have a least a smattering of it."[5] In the eighteenth century, many books, periodicals, works of art, and so on included scientific themes designed to meet this desire. In addition, shops, fairs, private lecture halls, and other public locations created spaces for scientific appropriation and purchase. I explore the opportunities for gaining access to scientific information in eighteenth-century France, with a particular focus on Paris. In addition, I examine the use to which people put the science they learned. In this milieu the public could approve or disapprove of science through their pocketbooks and their presence at scientific events. I attempt to elucidate the wealth of scientific activity that took place at the popular level within the French capital, decipher what the appropriation of science meant to the people involved, and explore how those individuals made use of their knowledge of popular science to help shape public opinion.

Science has long held a key place in studies of the Enlightenment. Earlier books by Ernst Cassirer, Carl Becker, Paul Hazard, Daniel Mornet, and, later, Peter Gay considered science crucial to the development of that movement as well as to the self-definition of its participants.[6] The great savants of the earlier period, and especially Isaac Newton, provided ideal role models for the thinkers of the eighteenth century who hoped to accomplish in other fields, both scientific and social, what Newton had done for physics. The centrality of science encouraged the application of its methods to problems, even if people often misunderstood the original scientific impetus. As a result, many eighteenth-century thinkers at least claimed to work from scientific principles, although this was not always the case. More recent histories of the Enlightenment, like those of Robert Darnton, Dena Goodman, and Roger Chartier, emphasize its social and cultural aspects rather than its intellectual development and have placed less emphasis on scientific matters. Instead, these scholars examine the transmission of knowledge and pursue questions regarding the circulation and diffusion of enlightened ideas, the development of enlightened sociability, and the institutionalization of the Republic of Letters in salons, freemason lodges, and academies.[7] In this book, I reassert the importance of science for understanding enlightened culture in the public sphere.[8]

I emphasize that aspect of the Enlightenment that focused on the utility of knowledge and the desire to disseminate information to a large number of people for the general betterment of society. For the most part, the philosophes did not address their ideas to peasants or even the working classes;[9] however, disseminators of enlightened ideas did target many members of urban society as the potential recipients of their efforts. While scholars analyze this phenomenon as seen through the readership of books and affiliation with enlightened institutions, I explore it through the appropriation of science, in the form of public lecture courses, participation in public science, and case studies demonstrating how people used their newly acquired knowledge. Clearly, then, the culture of Enlightenment science expanded and spread throughout the urban world; but, this leads to the question of exactly why eighteenth-century science interested people.

Historians of science have analyzed the place of science in public and the culture of science more generally. Works by Paula Findlen, Jan Golinski, and Larry Stewart have helped redefine the historiography for eighteenth-century science through an emphasis on the cultural impact and importance of scientific work.[10] For France, Emma Spary, Mary Terrall, Jessica Riskin, and Louise Robbins have reconceptualized the way we see science and the very culture of scientific activity. Terrall and Spary have explored the culture of scientific practice, Terrall for Maupertuis and the creation of one man's public image and Spary for the people working at the Jardin du Roi. Robbins, on the other hand, has outlined the vibrant commercial and literary culture surrounding exotic animals in Enlightenment Paris.[11]

Popular science in France has received some scholarly attention. Robert Darnton's study of animal magnetism includes an overview of popular scientific activities during the 1780s that touches on many forms of popular science including balloons and public lecture courses. In addition, Charles C. Gillispie wrote a detailed study of the ballooning efforts of the Montgolfier brothers from 1783 to 1784. More recently, Geoffrey Sutton wrote on the dissemination of science to members of the upper classes in the seventeenth and early eighteenth centuries. While these books discuss important issues, they do not fully analyze the place of science in French urban culture over the course of the eighteenth century.[12] French disseminators of science did not develop the idea of garnering public support for their efforts on their own. British practitioners of popular science, in particular, appeared by the end of the seventeenth century and established themselves firmly in the scientific culture of Great Britain by the early eighteenth century, a topic discussed by Larry Stewart among others. In addition, Oliver Hochadel has studied this phenomenon in German-speaking areas.[13]

Our understanding of the eighteenth-century urban world more generally, and the growth of public opinion and a consumer society, has expanded considerably over the last quarter century thanks to the work of historians such as Daniel Roche, David Garrioch, Colin Jones, and Arlette Farge. Garrioch successfully examines the development of Paris at street level and offers modern readers a Mercier-style understanding of how the city operated. Concerning the role of science Garrioch makes the point, although he only briefly elaborates on it, that enlightened culture in Paris "was increasingly 'scientific' and prided itself on being 'rational.'"[14] Analyses of how Parisians consumed products have proliferated. Colin Jones argues strongly for an increase in consumerism over the course of the Enlightenment, an occurrence he ties to the growth of a developing bourgeois class.[15] Daniel Roche, for his part, has written eloquently on consumer culture. A focus of his richly detailed and revealing work centers on the dynamic nature of eighteenth-century consumerism. He examines the consumption of books, clothes, and other material possessions, and portrays the eighteenth century as one in which the people of France learned how to consume even as more and more products became available to them.[16]

While Roche tends to concentrate on the circulation of goods, I examine the commodification of science through public lectures, subscriptions, and club memberships. This is a movement that engaged a large group of consumers keen to participate in enlightened activities, but perhaps unwilling, or unable, to acquire the necessary knowledge through more advanced scientific treatises replete with mathematics. The result is a somewhat broader sphere of activity in terms of what constituted the Enlightenment, the arenas in which people gained access to it, and what they did with it once they got it. This study, therefore, expands our understanding of the growth of the enlightened public sphere and the development of consumer culture through the lens of popular science, and offers insight into the interconnections between science, urban culture, and the Enlightenment.[17]

Popular science took many forms in the eighteenth century. Art, books, periodicals, pamphlets, plays, and poems, for example, could focus on, or at least refer to, scientific activities.[18] In particular, references to science in written works abounded during the eighteenth century. Everyone tried to demonstrate his or her knowledge of the sciences through occasional references to debates, ideas, formulas, and so on. Books from every genre, including novels, treatises on philosophy, letters, and even pornography, might include scientific references. Right in the middle of Denis Diderot's licentious novel *The Indiscreet Jewels*, for example, a debate breaks out between Cartesians and Newtonians. Jean-Jacques Rousseau claimed in *Julie, or The New Heloise* that "science is a currency much in demand" although he went on to criticize scientific popularizers as individuals more interested in having an audience than in educating people.[19]

Science found a home in every part of France and, in particular, in Paris. Students in the colleges and at the university could receive instruction in natural philosophy and medicine. In addition, academies offered a state-sponsored institutional home for the practice of science. Savants affiliated with the chief scientific academy, the Académie Royale des Sciences, even received a stipend although most found it necessary to supplement these funds through other means.[20] Individuals gained membership in this society, as in many of the national academies, only after an intense scrutiny from current academicians, their patrons, and state officials. Parisians also had the Jardin du Roi, a center for the study of natural history that hosted lectures in subjects like anatomy.[21] Some scientific work took place in more industrial settings with specifically practical goals in mind. Antoine-Laurent Lavoisier, for example, undertook improvements to gunpowder as the commissioner of the Régie des Poudres et Salpêtres, located at the Arsenal where he established a laboratory. The general public knew of his chemical work and appropriated it into their understanding of the world. Nikolas Karamzin, a Russian traveler in Paris, claimed Lavoisier's influence was so great, "for some years young beauties have loved to explain the tender agitations of their hearts in terms of chemical processes."[22] In addition, science appeared during the annual Paris fairs and even on the streets. "Our boulevards," claimed the commentator Jean-Baptiste Pujoulx, "have become schools of physics."[23] The increasing difficulty in getting state or noble patronage helps explain the decision made by popularizers of science to turn to the general public for support.

I consider the nature of popular science itself and its relationship to both urban culture and elite science. A number of different models exist for understanding the practice of popularization.[24] Questions of who the target audience for scientific dissemination should be, who should do the popularizing, and why it should be undertaken have led to a wide range of responses. In the eighteenth century, for example, elite savants might diffuse science to lay people. This model, however, often assumes a relatively simplistic top-down approach as well as the passivity of the audience. The examples of divining rods and balloons, discussed in this book, demonstrate the energetic role assumed by the public in scientific endeavors, in some cases even when savants wished they would stay away. Thus, people actively appropriated science in general, and even demanded specific kinds of science to meet their own cultural needs rather than the interests of the savants.

On entering the public sphere, the audience for popular science influenced, through their purchasing power, the direction, form, and manner in which science reached them. It is a mistake, then, to see popular science as simply an opportunity for the dissemination of elite science to a nonelite audience. The gradual development of the taste for science required the participation of multiple groups from different levels all of whom had unique goals and aspirations. This process was explicitly connected with the growth of consumption during eighteenth-century France. In this case, individuals purchased access to knowledge as well as entertainment. This aim on the part of the audience was not always recognized as legitimate on the part of the popularizers, some of whom hoped that their work would facilitate the creation of an enlightened elite rather than an audience concerned chiefly about cultural status. In the dissemination of science and its appropriation into urban culture, philosophes, popularizers, and audiences all drew on a set of shared ideas that each individual understood and appropriated in different ways and for distinct purposes. Thus, not only did elite scientific ideas find their way into the public sphere, but the efforts of amateurs and popularizers also influenced and infiltrated the cultural sphere of the social and scientific elites.[25]

Elite savants and philosophes often popularized science in order to persuade their peers regarding the efficacy of a particular scientific theory, program, or method. However, a form of popular science also grew within urban, enlightened culture that led to the propagation of a myriad of theories, ideas, and concepts. This kind of scientific activity developed due to the growing interest on the part of some popularizers to forge an occupation out of selling science to a broader audience combined with a rising interest among the urban public for information about science that was devoid, relatively, of polemics. An industry centered on scientific popularization developed as a consequence of public interest. The practice of popularization fed on both the desire of members of the Republic of Letters to disseminate ideas as well as on the increasing enthusiasm among certain members of the general population to understand scientific matters. As a result, a system of popular science arose that sought simultaneously to meet the propagandistic needs of the savants and the educational concerns of the people. Certainly, the desire for popularized science far outweighed the ability or interest of philosophes to provide it despite the belief of some of them that science should be popularized. Diderot, for example, suggested that people should "hasten to popularize [natural] philosophy."[26] Thus, in order for science to fully enter the public sphere, a whole new group of people had to take on the practice of popularization.

These disseminators of popular science found themselves in a tricky position. They relied on the good graces of their more elite colleagues as well as the financial support of the general population for their livelihood. In order to be successful and, as some of them wished, obtain the status of an elite savant, popularizers had to please both their clientele and their more elite brethren. Arguably, the need to please their audience played the most crucial part in their day-to-day existence, as they needed a steady stream of customers in order to make a living. This led popularizers to seek out the largest audience possible and made them fairly receptive to their needs and opinions. In other words, they catered their scientific work to meet the wants and expectations of their students. As such, popular science did not always look much like the science from whence it came; this is apparent in the work of Perrin and his savant dog. A similar conflict confronted proponents of the literary Enlightenment who applauded the growth of literacy and the development of a market for literature, while at the same time these proponents decried the growth of the novel as singularly unenlightened.[27] Nonetheless, popularizers functioned as cultural intermediaries between multiple aspects of scientific life. They funneled science to a broader portion of the general population than it otherwise would have reached. The desire of people to learn about science, regardless of whether or not their choice of venues matched the wishes of elite savants, formed a significant aspect of the culture of Enlightenment.

While the philosophes felt that science was crucially important for the Enlightenment and targeted it as a subject area that everybody should know something about, few people could pick up and learn about science on their own. Thus, they sought presentations and courses ranging from a single lecture to weeks-long explorations. To further complicate the process, eighteenth-century science at all levels existed in a state of flux as savants continually developed new categories, methods, and topics over the course of the century. Popular science existed within this matrix of developing, professionalizing, and transforming scientific disciplines.

The practice of popularization took place in nearly every discipline during the eighteenth century and public lecture courses appeared on almost every topic. A cursory examination of a single journal from the second half of the eighteenth century reveals the number and variety of public lecture courses being offered. In 1765, for example, the *Affiches de Paris* advertised classes on general mathematics, experimental physics, natural history, botany, geometry, chemistry, and pharmacy. This newspaper also promoted numerous nonscientific classes offered on topics like commerce, history, geography, art, architecture, and foreign languages. By the 1780s, opportunities for satisfying intellectual needs and acquiring cultural capital had risen dramatically in overall number and variety. Now people could attend classes on navigation, physical astronomy, electricity, French writing and pronunciation, the properties of air, vocal and instrumental music, midwifery, surgery, Latin, the vegetable kingdom, the structure of the eye, the painting and sculpture of animals, Hebrew, poetry, optics, mineralogy, and agriculture. These courses did not necessarily offer a watered-down version of real science, although scientific material was frequently stripped of mathematical calculations. On the other hand, auditors taking M. Dupont's mathematics class in 1779, for example, studied such advanced subjects as differential and integral calculus. It seems unlikely that many people today, after already finishing their schooling, would voluntarily take a calculus class; but Dupont's classes were perennial favorites in Paris for over a decade.[28]

Rather than grappling with all scientific disciplines, one representative area will provide the focus for this book – namely experimental physics. This branch of science highlights several important themes. First, although suggested as a course of study by Bacon in the seventeenth century, it developed as a distinct branch of science only during the eighteenth century. In other words, experimental physics evolved in juxtaposition with the Enlightenment and, thus, offers the modern historian an opportunity to see how members of the public sphere shaped it for their own purposes without having to first eradicate old perceptions of its use. Experimental physics also serves as a good entry into the study of popular science and the Enlightenment based on its direct connections with scientific icons like Isaac Newton. Although calculus would also make its way across the English Channel and become an object of popularization, the experimental side of science, like that expounded in Newton's *Opticks*, found an immediate home in France through its specific association with rationalism and the origins of the Enlightenment. Thus, for the French, experimental physics exemplified the project of the Enlightenment and afforded an accessible path for amateurs to participate in and gain access to the scientific culture of the eighteenth century.[29]

The scope of experimental physics, broadly defined by contemporaries, also allows for a wide range of fields to be included in this study. John Heilbron has derived a definition of experimental physics from eighteenth-century sources that will be appropriated here. Experimental physics, Heilbron writes, is a body of knowledge that was "popular, authoritative, and influential. It should draw on Newton but miss his subtlety, and utilize continental work without regard to doctrine. It should illustrate theory with experiment whenever possible and use numbers for measurement but not for calculation. As for coverage, it must omit the biological sciences and attend to, at a minimum, mechanics, optics, fluid mechanics, pneumatics, heat, meteorology, geophysics, electricity, and magnetism."[30] Experimental physics includes many aspects of natural philosophy in general, the key being the use of experiments rather than mathematics as the primary method of demonstration. Thus, while this discipline holds a specific place in eighteenth-century thought, at the same time it is broad enough to allow for generalizations to be made concerning the overall significance of popularization.

Experimental physics was also the science most influenced by trend and fashion during the course of the eighteenth century. It monopolized the French popular consciousness during the Enlightenment through a mixture of good timing and spectacular presentations. Experiments before the royal court, like Nollet's electric shocks, and those before entire cities, such as the balloon flights of Jean-François Pilâtre de Rozier and Jacques-Alexandre-César Charles, took science out of the laboratory and made it a public commodity.

While books, periodicals, universities, and academies all provided a breadth of scientific popularization at different levels and for different audiences, this book focuses on popular science within urban culture more generally. More than ever before, public lectures and demonstrations, clubs, and other activities arose in the eighteenth century as new opportunities for the general population to gain access to and appropriate science. These arenas for popular science were not restricted to people of a certain education. In fact, popular science, and public lecture courses in particular, was often set at a level that could be understood by pretty much anyone; this was a bone of contention between popularizers and their critics who felt that in some cases popular science lacked any sort of real scientific content. In reality, some popularizers had specific theoretical content in mind for their courses while others were admittedly more interested in theatrics. Chapter 2 discusses this eclectic group and the breadth of their dissemination activities, and describes the points of interaction between popular science and the public. Thus, this chapter examines popularizers as well as the practice of scientific popularization. The individuals who gravitated towards this profession came from a multitude of backgrounds, educational experiences, and abilities, and they approached this job with a vast array of goals in mind. This chapter explores the composition of this group, the methods they used to popularize science, and the topics they discussed.

Identifying the audience, cost, and location of popular science helps reveal its place in urban culture. Chapter 3 looks at the audience, identified through advertisements and course descriptions, as well as the economics of courses. Advertisements outlined the content and length of a course, the details of when and where it would be offered, and the price structure. They often alluded to the perceived audience for the course, with some notices specifying exactly who the popularizers thought should attend. I use this information to determine a range of potential auditors and the cost of acquiring popular science. In addition, I map out the topographical details of popularization and sketch the geography of dissemination in Paris. While popular science could be found all over the city in one form or another, popularizers concentrated their activities in specific locations depending on their target audience. The geography of popular science changed over time and, as it did, individual disseminators forged connections between social groups, crossed boundaries, and addressed different audiences.

Some individuals formalized public lecture courses in the 1770s and 1780s through the creation of clubs and organizations, firmly situated within the public sphere. As discussed in Chapter 4, organizations like the Musée de Monsieur drew in members with a variety of motivations. Some desired the status accorded those who belonged to this club. Others wanted to make use of the library and experimental equipment and, most important to my purposes, to attend the different lecture courses offered on an almost daily basis. These clubs proliferated during the last quarter of the eighteenth century and some of them survived well into the nineteenth century.

The spread of popular science in urban culture encouraged people to speak about science and participate in scientific debates, whether or not they really were qualified to do so. While the public interest in Mesmerism has received considerable attention, a less well-known debate over the utility of divining rods also took place in the eighteenth century. Chapter 5 looks at this public debate beginning first with a case from the 1690s. This earlier case is juxtaposed with a case from the 1770s and 1780s to demonstrate the changing nature of the participants in the quarrel as well as the different methods used to analyze the question. In the end, the acquisition of scientific cultural capital affected the willingness of the public, both in reality and in the minds of those individuals trying to control and influence public opinion, to participate in scientific debates.

Finally, Chapter 6 examines hot air and hydrogen balloons, first invented in France in 1783. They quickly spread to become a popular phenomenon. The discovery of balloons spawned reams of poetry, plays, and scientific treatises, as well as much admiration. Balloon launches, many performed by individuals who also offered courses in experimental physics, drew crowds of up to 100,000 people, many of whom both paid for the privilege of attending and validated the experiment performed. Such public venues and popular approbation made ballooning the popular science *par excellence* and an early example of mass culture. With the advent of ballooning, popular science soared to new heights, literally and figuratively, and established itself as a crucial component of eighteenth-century urban life.

Notes

1 Mercier, *Panorama of Paris*, ed. Jeremy Popkin (University Park, PA: Pennsylvania State University Press, 1999), 30.

2 Voltaire to Pierre Robert le Cornier de Cideville, [16 April] 1735, in Voltaire, *Correspondance*, ed. Theodore Besterman, 107 vols. (Geneva: Institut et Musée Voltaire, 1965), LXXXVII:132.

3 For an account of Nollet's experiments see Nollet, "Observations sur quelques nouveaux phénomènes d'Electricité," *Mémoires de l'Académie Royale des Sciences* (1746): 1–23 and Nollet, "Sur l'électricité," *Histoire de l'Académie Royale des Sciences* (1746): 1–17. Nollet was not the first to try this experiment; Louis-Guillaume le Monnier had shocked a chain of 140 courtiers in a similar fashion. See John Heilbron, *Electricity in the Seventeenth and Eighteenth Centuries: A Study of Early Modern Physics* (Berkeley: University of California Press, 1979), 318–20. For the castrati see Joseph Aignon Sigaud de la Fond, *Précis historique et expérimentale des phénoménes électriques depuis l'origine de cette découverte jusqu'à ce jour* (Paris: Rue et Hôtel Serpente, 1781), 284–91; and *Journal encyclopédique*, July 1772, 126–9.

4 *Journal de Paris*, 12 October 1784, 1208; also see the handbill issued by Perrin, *Amusemens de physique* (Paris, P. de Lormel, 1787b).

5 Jean Ferapie Dufieu, *Manuel physique, ou manière courte et facile d'expliquer les phénomènes de la nature* (Paris: J.-T. Herissant, 1758), vii–viii.

6 Ernst Cassirer, *The Philosophy of the Enlightenment*, trans. Fritz C.A. Koelln and James P. Pettegrove (Princeton: Princeton University Press, 1951); Carl L. Becker, *The Heavenly City of the Eighteenth-Century Philosophers* (New Haven: Yale University Press, 1932); Paul Hazard, *The European Mind: The Critical Years, 1680–1715*, trans. J. Lewis May (New York: Fordham University Press, 1990); Hazard, *European Thought in the Eighteenth Century: From Montesquieu to Lessing*, trans. J. Lewis May (Glouster, MA: Peter Smith, 1973); Daniel Mornet, *Les Origines intellectuelles de la Révolution Française, 1715–1787* (Paris: Armand Colin, 1967 [1933]); and Peter Gay, *The Enlightenment: An Interpretation*, 2 vols. (New York: Norton, 1966–1969).

7 Robert Darnton explored scientific themes in his earlier book *Mesmerism and the End of the Enlightenment in France* (Cambridge, MA: Harvard University Press, 1968), but has not focused on this topic in his subsequent works like *The Literary Underground of the Old Regime* (Cambridge, MA: Harvard University Press, 1982) and *The Forbidden Best-Sellers of Pre-Revolutionary France* (New York: Norton, 1996). Chartier, *The Cultural Origins of the French Revolution*, trans. Lydia G. Cochrane (Durham NC: Duke University Press, 1991); Goodman, *The Republic of Letters: A Cultural History of the French Enlightenment* (Ithaca: Cornell University Press, 1994). Also see Margaret Jacob, *Living the Enlightenment: Freemasonry and Politics in Eighteenth-Century Europe* (Oxford: Oxford University Press, 1991); Roger Hahn, *The Anatomy of a Scientific Institution: The Paris Academy of Sciences, 1666–1803* (Berkeley: University of California Press, 1971); and Daniel Gordon, *Citizens without Sovereignty: Equality and Sociability in French Thought, 1670–1789* (Princeton: Princeton University Press, 1994).

8 On the public sphere see Jürgen Habermas, *The Structural Transformation of the Bourgeois Public Sphere*, trans. Thomas Burger (Cambridge, MA: MIT Press, 1989); Thomas H. Broman, "The Habermasian Public Sphere and 'Science *in* the Enlightenment,'" *History of Science* 36 (1998): 123–49; and Broman and Lynn K. Nyhart (eds.). "Science and Civil Society", *Osiris* 17 (2002). Also see Bernadette Bensaude-Vincent *L'opinion publique et la science: A chacun son ignorance* (Paris: Institut d'édition Sanofi-Synthélabo, 2000); James Van Horn Melton, *The Rise of the Public in Enlightenment Europe* (Cambridge: Cambridge University Press, 2001); and

T.C.W. Blanning, *The Culture of Power and the Power of Culture: Old Regime Europe, 1660–1789* (Oxford: Oxford University Press, 2002).

9 Harvey Chisick, *The Limits of Reform in the Enlightenment* (Princeton: Princeton University Press, 1981).

10 Paula Findlen, *Possessing Nature: Museums, Collecting, and Scientific Culture in Early Modern Italy* (Berkeley: University of California Press, 1994); Jan Golinski, *Science as Public Culture: Chemistry and Enlightenment in Britain, 1760–1820* (Cambridge: Cambridge University Press, 1992); Larry Stewart, *The Rise of Public Science: Rhetoric, Technology, and Natural Philosophy in Newtonian Britain, 1660–1750* (Cambridge: Cambridge University Press, 1992).

11 Emma Spary, *Utopia's Garden: French Natural History from Old Regime to Revolution* (Chicago: University of Chicago Press, 2000); Mary Terrall, *The Man Who Flattened the Earth: Maupertuis and the Sciences in the Enlightenment* (Chicago: University of Chicago Press, 2002); Jessica Riskin, *Science in the Age of Sensibility: The Sentimental Empiricists of the French Enlightenment* (Chicago: University of Chicago Press, 2002); Louise E. Robbins, *Elephant Slaves and Pampered Parrots: Exotic Animals in Eighteenth-Century Paris* (Baltimore: Johns Hopkins University Press, 2002). More generally see Roy Porter, ed., *The Cambridge History of Science*, Vol. 4, *Eighteenth-Century Science* (Cambridge: Cambridge University Press, 2003); and William Clark, Jan Golinski, and Simon Schaffer, eds., *The Sciences in Enlightened Europe* (Chicago: University of Chicago Press, 1999).

12 Darnton, *Mesmerism*; Charles C. Gillispie, *The Montgolfier Brothers and the Invention of Aviation: 1783–1784* (Princeton: Princeton University Press, 1983); and Geoffrey Sutton, *Science for a Polite Society: Gender, Culture, and the Demonstration of Enlightenment* (Boulder, CO: Westview Press, 1995).

13 For Great Britain see Simon Schaffer, "The Consuming Flame: Electrical Showmen and Tory Mystics in the World of Goods," in *Consumption and the World of Goods*, eds. John Brewer and Roy Porter (London: Routledge, 1993): 489–526; Schaffer, "Natural Philosophy and Public Spectacle in the Eighteenth Century," *History of Science*; Golinski, *Science as Public Culture*. For German-speaking areas see Oliver Hochadel, *Öffentliche Wissenschaft: Elektrizität in der deutschen Aufklärung* (Göttingen: Wallstein, 2003).

14 David Garrioch, *The Making of Revolutionary Paris* (Berkeley: University of California Press, 2002), 265; also see his *The Formation of the Parisian Bourgeoisie, 1690–1830* (Cambridge, MA: Harvard University Press, 1996). Arlette Farge, *Subversive Words: Public Opinion in Eighteenth-Century France*, trans. Rosemary Morris (University Park, PA: Pennsylvania State University Press, 1994).

15 Colin Jones, "The Great Chain of Buying: Medical Advertisement, the Bourgeois Public Sphere and the Origins of the French Revolution," *American Historical Review* 101 (1996): 13–40. On the idea of the bourgeoisie in eighteenth-century France see Sarah Maza, *The Myth of the French Bourgeoisie: An Essay on the Social Imaginary, 1750–1850* (Cambridge, MA: Harvard University Press, 2003), esp. chps. 1–3.

16 Daniel Roche, *The People of Paris: An Essay in Popular Culture in the Eighteenth Century*, trans. Marie Evans with Gwynne Lewis (Berkeley: University of California Press, 1987); Roche, *The Culture of Clothing: Dress and Fashion in the "Ancien Régime"*, trans. Jean Birrell (Cambridge: Cambridge University Press, 1994); Roche, *A History of Everyday Things: The Birth of Consumption in France, 1600–1800*, trans. Brian Pearce (Cambridge: Cambridge University Press, 2000); and Roche, *La France des lumières* (Paris: Fayard, 1993).

17 On the growth of consumption see John Brewer and Roy Porter, eds., *Consumption and the World of Goods* (London: Routledge, 1993); Neil McKendrick, John Brewer, and J.H. Plumb, *The Birth of a Consumer Society: The Commercialization of Eighteenth-Century England* (Bloomington: Indiana University Press, 1985). On the commercialization of English science see e.g. Larry Stewart, "Public Lectures and Private Patronage in Newtonian England," *Isis* 77 (1986): 47–58; Stewart, "The Selling of Newton: Science and Technology in Early Eighteenth-Century England," *Journal of British Studies* 25 (1986): 178–92; Patricia Fara, *Sympathetic Attractions: Magnetic Practices, Beliefs, and Symbolism in Eighteenth-Century*

England (Princeton: Princeton University Press, 1996), esp. chp. 2.

18 On art, see Barbara Stafford, *Artful Science: Enlightenment Entertainment and the Eclipse of Visual Education* (Cambridge, MA: MIT Press, 1994). On newspapers, see Jean Sgard, ed., *Dictionnaire des journaux, 1600–1789*, 2 vols. (Paris: Universitas, 1991); Sgard, *La presse provinciale au XVIII*[e] *siècle* (Grenoble: Centre de Recherches sur les Sensibilités, 1983); and Gilles Feyel, "Médecins, empiriques et charlatans dans la presse provinciale à la fin du XVIIIe siècle," in *Le Corps et la santé: Actes du 110*[e] *congrès national des sociétés savantes* (Paris: C.T.H.S., 1985): 79–100.

19 Diderot, "The Indiscreet Jewels." Trans. Sophie Hawkes, in *The Libertine Reader: Eroticism and Enlightenment in Eighteenth-Century France*, ed. Michel Feher (New York: Zone, 1997): 333–541; for Diderot's discussion of the debate between Cartesians and Newtonians see 365–6. Rousseau, *Julie, or the New Heloise*, trans. Philip Stewart and Jean Vaché (Hanover NH and London: University Press of New England, 1997), 46.

20 Significantly, even elite savants, members of an academy or with state patronage, required multiple sources of income in order to make a living. On the finances of academicians see David J. Sturdy, *Science and Social Status: The Members of the Académie des Sciences, 1666–1750* (New York: Boydell Press, 1995).

21 On education see L.W.B. Brockliss, *French Higher Education in the Seventeenth and Eighteenth Centuries: A Cultural History* (Oxford: Clarendon, 1987); on the Jardin du Roi, and public lectures, see Anita Guerrini, "Duverney's Skeletons," *Isis* 94 (2003): 577–603, esp. 587–91; Yves Laissus, "Les cabinets d'histoire naturelle," in *Enseignement et diffusion des sciences en France au XVIII*[e] *siècle*, ed. René Taton (Paris: Hermann, 1964), 659–712; and Spary, *Utopia's Garden*.

22 Karamzin, *Letters of a Russian Traveler, 1789–1790*, trans. Florence Jonas (New York: Columbia University Press, 1957), 226–7.

23 Pujoulx, *Paris à la fin du XVIII*[e] *siècle* (Paris: Brigite Malthé, 1801), 33.

24 For a useful discussion of the historiography of popular science, see Roger Cooter and Stephen Pumfrey, "Separate Spheres and Public Places: Reflections on the History of Science Popularization and Science in Popular Culture," *History of Science* 32 (1994): 237–67. Also see Mary Fissell and Roger Cooter, "Exploring Natural Knowledge: Science and the Popular," in Porter, ed., *Eighteenth-Century Science*, 129–58.

25 Chartier, "Popular Appropriations: The Readers and Their Books," in *Forms and Meanings: Texts, Performances, and Audiences from Codex to Computer* (Philadelphia: University of Pennsylvania Press, 1995): 83–97; and also Chartier, "Intellectual History and the History of Mentalités: A Dual Re-evaluation," in his *Cultural History: Between Practices and Representations*, trans. Lydia G. Cochrane (Ithaca: Cornell University Press, 1988): 19–52.

26 Diderot, "Thoughts on the Interpretation of Nature" in *Thoughts on the Interpretation of Nature and Other Philosophical Works*, ed. David Adams (Manchester: Clinamen Press, 1999), 59.

27 See Melton, *Rise of the Public in Enlightenment Europe*, 94–6.

28 In 1785, more than 100 public lecture courses were advertised in the *Affiches de Paris* alone. For Dupont, see *Journal de Paris*, 3 November 1779, 1253. For a brief analysis of public lecture courses in France see Reed Benhamou, "Cours publics: Elective Education in the Eighteenth Century," *Studies on Voltaire and the Eighteenth Century* 241 (1986): 365–76.

29 Thomas S. Kuhn, "Mathematical versus Experimental Tradition in the Development of Physical Science," in *The Essential Tension* (Chicago: The University of Chicago Press, 1977): 31–65.

30 Heilbron, "Experimental Natural Philosophy," in *The Ferment of Knowledge: Studies in the Historiography of Eighteenth-Century Science*, eds. G.S. Rousseau and Roy Porter (Cambridge: Cambridge University Press, 1980), 363.

2

The practice of popularization

The eighteenth century witnessed a growing public interest in natural philosophy. As one savant wrote, "physical phenomena gave [people] an indescribable pleasure."[1] The practice of popularization formed an integral part of eighteenth-century urban culture and illustrates one instance of how individuals could gain access to science. In fact, the steady rise of scientific dissemination in all its forms constituted a significant part of that aspect of the Enlightenment, along with literary projects like the *Encyclopédie* of Diderot and d'Alembert, which worked diligently to make information available to the greatest number of people. For their part, popularizers typically focused on the dissemination of science, rather than on the development of new theories. Disseminators of experimental physicists concentrated on both the usefulness and the entertainment value of science. They stressed the utility of knowledge, in this case natural philosophy, and claimed that it should be made accessible to the greatest number of people for the general benefit of society.

Mid-level savants, for the most part, composed the group that performed this task. These popularizers employed a variety of oral and visual techniques in which the audience watched science in action and sometimes actually performed or participated in the experiment. When middling savants published popular science books, as they sometimes did, they often did so by compiling lectures they had previously delivered in their courses. Popularizers intended their audience to read about and, if possible, to duplicate the demonstrations described therein. Since popularizers usually did not concern themselves with persuading people to accept a specific theory, the door remained open for the audience to appropriate ideas and experiments in a truly eclectic fashion. In addition, middling savants forged a new relationship with the public. They reacted to the wants and needs of their audience, a group larger, and more socially diverse, and at a more varied level of education, than most philosophes had ever tried or wanted to address. This audience supported the popularizers socially and economically in their efforts to make a living for themselves in the commercial dissemination of science. Among other jobs, middling savants worked as educators, entrepreneurs, and disseminators.

The background for popular science, 1660–1740

At the beginning of the eighteenth century, Pierre Polinière, in an experiment designed for students attending one of the Parisian colleges, placed an overripe, shriveled apple under a bell jar and pumped the air from the jar, as a result of which the apple first smoothed out and then exploded. His audiences wildly and widely applauded him for this demonstration of air pressure. Polinière used this experiment to exhibit the internal force of expanding air in the apple and concluded, among other things, that seeds grow because the heat of the sun expands the air inside them. Whatever Polinière intended the audience to take from this experiment, exploding fruit captured the imagination of people interested in science as well as those individuals who liked a good show. Over the course of the seventeenth century, popularization activities like Polinière's steadily grew as savants became more and more interested in reaching a lay audience and, likewise, as the urban population began to express interest in scientific affairs.[2]

Popularization during the age of the Enlightenment capitalized on an existing trend created by savants and educators in the seventeenth and early eighteenth centuries. This earlier movement peaked in the period between 1690 and 1740 when savants actually began turning to the general public for support in scientific debates like the one over divining rods discussed in Chapter 7 of this book. Questions of scientific legitimacy and authority were debated in journals and popular pamphlets written explicitly to sway learned public opinion. At the same time, philosophes such as Voltaire and Algarotti began to publish works with an eye towards persuading their peers that Newtonian theories should prevail over their Cartesian counterparts. Thus, there existed a long-term shift towards popularizing science for interested individuals as well as a turn towards involving the general public in scientific debates.

The growth of popular science also drew inspiration from the more general development of a culture of scientific practice. The period from 1650 to 1750 saw the creation of numerous state-sponsored professional organizations, both in France and around Europe, such as the Académie Royale des Sciences, and the establishment of various publications, such as the *Journal des savants* and the *Journal de Trévoux*, which encouraged the consumption of science by the general public. These institutions provided the basis for the appropriation of science by a wider group of people than previously had been possible. In addition, a number of private organizations appeared where elite savants could mix with amateurs and middling savants. These groups, such as the Montmor Academy and the weekly conferences sponsored by Théophraste Renaudot, offered both a model for and an alternative to the state-controlled Académie Royale des Sciences.[3] Private academies created a forum where non-academicians could meet with other interested individuals to exchange findings and conclusions. This new audience did not necessarily have an innate interest in scientific discoveries per se, but rather a keen interest in useful and entertaining science. Thus, over time, a group of savants began to alter their writings in order to limit the time spent on mathematical proofs and theoretical discussions and maximize the number of demonstrations and experiments. At the same time, other authors began addressing their audiences through literary works. Consequently, the size of the potential audience grew quickly during this period.

In addition to the rapid expansion of the audience for science, this period also witnessed a change in both the manner and the location of popular science. In order to facilitate a public interest in science, disseminators established new, more accessible forums and developed new pedagogical methods that displayed science in a more engaging fashion. Simultaneously, a shift occurred in the kind of science being popularized, with visual, non-mathematical sciences stepping up to the forefront. Popular science, therefore, came to form a part of everyday culture for many people and it became appropriate and desirable for members of society to be able to speak about science.[4]

While some savants established academies and wrote books or contributed to periodicals, other savants set the stage more directly for the future development of public lecture courses. Some disseminators took the science out of books or formal discourses and carried it into different social arenas. In particular, educational institutions and private salons became places where science reached an even broader audience than ever before. Individuals interested in studying science found that they could learn about it without considerable formal education or even literacy; they needed only time and money. Two savants in particular exemplify this new educational methodology, namely Jacques Rohault and Pierre Polinière. They had a remarkable facility for providing concrete demonstrations of natural philosophy. This encouraged the dissemination of scientific ideas and laid the foundation for the growing trend in the consumption of popular science.

Rohault performed his work in a salon setting. He had started his career as a mathematics tutor where his skill and reputation earned him the patronage of Claude Clerselier, a friend of René Descartes and later editor of Descartes's correspondence. Clerselier converted Rohault to Cartesianism and Rohault, in turn, married Clerselier's daughter Geneviève, thus sealing their relationship. Clerselier, for his part, hoped that Rohault would "spread the truths that

Descartes had taught."[5] In any case, eighteenth-century savants credited Rohault with the introduction of "reason and experimentation" into the teaching of physics.[6] His most important contribution to the dissemination of scientific thought lay in his initiation of regular meetings, held in his home, concerning physics. These "Cartesian Wednesdays" began as early as 1659 and continued up until his death in 1672 at which point his student, Pierre Sylvain Régis, took over.[7]

Rohault's lectures held great appeal, and a fairly large number of people attended them at various times including, on several occasions, Christiaan Huygens. More importantly, however, women formed a significant portion of the audience. Every Wednesday people of "all ages, both sexes, and all professions" would meet to watch Rohault as he gave demonstrations intended to provide an experimental base for Descartes's physics. To accomplish this task, Rohault accumulated a large collection of scientific instruments and developed numerous experiments. Even academicians regarded his efforts favorably and, although he never gained admission into the Académie Royale des Sciences, savants acknowledged that his work did a great deal to legitimize Cartesianism in France.[8]

Rohault, however, remained largely outside of official science. His work popularized Cartesianism, but he did not introduce any new ideas. His contribution to the history of popular science stems from three aspects of his approach. First, he spoke directly to amateurs, many of whom were women. Second, he lectured in a relatively informal setting. In other words, he took science out of the academy and presented it directly to his audience. Last, he provided physical demonstrations rather than mathematical proofs to illustrate the theories he discussed. This enabled individuals without a scientific background to follow the logic of the experiments and understand the results. Rohault truly established a new setting, a new audience, and a new methodology for popular science.[9]

In addition to Rohault's integration of popular science into polite society, the educational sphere also underwent a significant pedagogical change. Pierre Polinière and his experiments, with the exploding fruit mentioned earlier, best exemplify the efforts in this area. His work marks the beginning of a shift towards empiricism in the classrooms of Paris. Both in his teaching method and in his published experiments, Polinière changed the nature of physics education. Before Polinière, professors teaching physics relied almost exclusively on formal lecturing. Students only started to receive visual, experimental instruction at the turn of the eighteenth century when natural philosophy professors and their students called upon the services of individuals like Polinière to provide concrete demonstrations of physical principles. In some respects, Polinière took the first step in a pedagogical revolution.[10]

Polinière himself did not work for a specific college. He toiled as a freelance demonstrator of natural philosophy and mathematics, topics he had first studied at the University of Caen and later at the University of Paris. In the 1690s, he set himself up in Paris as a tutor. He continued his own work in mathematics and also became interested in the properties of phosphorus. Historians often credit him, along with Francis Hauksbee, with the discovery of electroluminenscence.[11] He taught physics so clearly that some students hired him to help them. This, in turn, led the faculty of philosophy at the Collège d'Harcourt to ask Polinière to present demonstrations in natural philosophy for both their own and other students enrolled at the University of Paris. Polinière agreed and offered "a complete, but also amusing, course of study [that] strongly appealed to the tastes of young men."[12] His course, which generally lasted about two months, was a huge success. In addition, the book he wrote based on his lectures, the explicitly Cartesian *Expériences de physique*, also enjoyed enormous popularity and went through multiple editions throughout the eighteenth century.[13]

Rohault and Polinière operated on the cusp linking the interested general public, the Académie Royale des Sciences, and University of Paris. Popular science, therefore, developed and grew tangential to or even outside the realm of official, academy-based, science. Even so, the efforts of popularizers set the pattern for the work to follow. The goal of teaching natural philosophy to a literate but unschooled public later became the focus of popularized science in the eighteenth century.

Popularizers, 1735–1793

From these somewhat modest origins, public lecture courses emerged to become one of the dominant forms of scientific popularization in eighteenth-century France. An examination of the number of popular science courses advertised in French newspapers alone reveals a wide range of people offering courses during this period aimed at an extremely broad audience.[14] In the period from 1735 to 1793 more than seventy individuals offered a variety of courses and demonstrations just in experimental physics. If we include other scientific disciplines, such as natural history, then the number of popular science disseminators begins to climb even more dramatically.[15] The practice of scientific popularization rose throughout the century, reaching its highest levels during the period from about 1775 up to the first years of the French Revolution. During that period an average of fifteen experimental physicists offered more than forty courses in Paris each year. Since each class could have as many as a hundred auditors, and some of the larger classes had as many as four or five hundred, it can be estimated that in any given year several thousand people attended these public lectures. The discovery of ballooning in 1783, a breakthrough that lent itself to stunning visual displays and popular appeal, dramatically contributed to the rapid rise in popular science at the end of the century. Ballooning triggered such an upsurge in the popular appetite for experimental physics that twenty-three professors of experimental physics offered eighty-one classes in 1785 alone.

The dissemination of science grew steadily throughout the century with only one minor lull during the period 1768 to 1776, when the number of individuals who offered classes fluctuated between just seven and nine, and only a few new individuals entered the popular science profession. This temporary dip ended in 1777, when seven new popularizers began advertising their classes and the total number of courses rose from eleven in 1776 to thirty-three in 1777. There are several possible explanations for this sudden increase, the most obvious being that 1777 saw the publication of the first French daily newspaper, the *Journal de Paris*. Many popularizers advertised their courses in this periodical and it is possible that some, already active prior to that date, began to advertise here for the first time. Alternatively, an explanation can be found in the political machinations that took place in France in the wake of the Seven Years War. The Jesuits, long under attack in France by philosophes and Gallicans, found themselves the target of much animosity. In 1762, the Parlement of Paris abolished the Society of Jesus in France and ordered its members to leave the country. At this time, the Jesuits operated more than one hundred colleges throughout France, many of which offered courses in natural philosophy. These colleges needed to find new faculty in order to accommodate all the students. This fact, coupled with the normally high turnover rate among college professors, led to a vast increase in the number of vacant teaching positions. Many young savants, who might have been content to bide their time teaching popular science while waiting for teaching positions, academy posts, or royal sinecures, gained immediate employment as educators in the former Jesuit colleges. The growth in the number of teaching jobs did not occur immediately after 1762. It took a few years for the different regions to comply with the edict and find alternative funds to pay the new teachers. By the late 1760s, however, the secularization of the Jesuit colleges was well under way.[16]

Over the course of the eighteenth century, the dissemination of science gradually emerged as a recognizable and legitimate career choice. A dearth of biographical data prevents a detailed prosopographical analysis of popularizers. However, the available information does allow for some generalizations. Many popularizers, for example, held membership in academies, although only rarely in the Académie Royale des Sciences in Paris. Instead, they tended to hold memberships in provincial or foreign academies. The academies of Montpellier, Angers, and Bavaria, for example, asked Joseph-Aignan Sigaud de la Fond to join.[17] Allard claimed membership in the academies in Auxerre and Angers.[18] Only Nollet, and after him Mathurin-Jacques Brisson, earned a place in the Académie Royale des Sciences, positions they held thanks to the considerable patronage they garnered from the crown and other academicians. Jacques-Alexandre-César Charles, however, did win a place in the Institut when it arose from the ashes of the disbanded Académie Royale des Sciences during the French Revolution.[19]

While popularizers tried to earn their living through their public lecture courses, they also might hold one or more additional jobs. Mathurin-Jacques Brisson, in addition to his position in the Académie Royale des Sciences, worked as a royal censor and an industrial inspector. Nollet, and after him Brisson, also held a teaching job at the Collège de Navarre. The monarch also helped fund the creation of Nollet's instrument collection, one of the finest in Europe at that time.[20] It seems that once Nollet gained an official position as an educator, it became possible for popularizers in general to make that shift. In the end, popularizers collected many teaching positions in order to enhance their status in society and earn enough money to pursue their scientific interests. Nollet, for example, also taught at the Ecole d'Artillerie and the Ecole Royale du Génie de Mézières. Allard was rewarded for his efforts as a popularizer with a position teaching mathematics at the Ecole Royale Militaire. Louis Lefèvre-Gineau held a newly renamed chair in experimental physics at the Collège Royal.

Some popularizers gained the attention and patronage of foreign heads of state. Nollet, for example, was well known throughout Europe and in 1739 Charles-Emmanuel III, the king of Sardinia, invited him to visit Turin and present some demonstrations for the royal court. Nollet also performed experiments for distinguished foreigners visiting Paris including the king of Denmark, Christian VII, in 1768. In the late 1770s Joseph II of Austria paid a visit to Ledru's *cabinet de physique*; he later claimed that he enjoyed this visit more than the "meaningless" demonstrations Lavoisier had performed for his benefit at the Académie Royale des Sciences. François Pelletier, somewhat more mysteriously, managed to obtain the patronage of the heir to the Spanish throne, Dom Gabriel.[21]

Aside from crowned heads of state, French nobles also gave their patronage to popularizers. Many nobles themselves had an amateur interest in the sciences and would often act as patrons to one or more savants. Ledru, for example, habitually performed experiments for the duc de Chartres, an amateur physicist who had once studied with Nollet. Pilâtre de Rozier gained the post of intendant of the physics, chemistry, and natural history cabinets in the household of the count and countess of Provence.[22] While patronage helped, popularizers also might work in other capacities. Jacques Bianchi, in addition to popularizing experimental physics, made and sold scientific equipment. In fact, it seems likely that his demonstrations grew out of a desire to display the machines and instruments in his shop. François Bienvenu also fits the model of a merchant and physicist.[23]

Savants sometimes supported each other. Pilâtre de Rozier, for example, got a job teaching chemistry at the Société Libre d'Emulation in Rheims thanks to the support of the academician Balthazar-Georges Sage. Nollet, when he first started working as a savant, landed a job as a laboratory assistant to Charles François de Cisternai Dufay and René-Antoine Réaumur. Through their support he gained membership to the Académie Royale des Sciences in 1739. Nollet, for his part, passed on many of his titles, jobs, and positions to his protégé, Brisson. The demonstrator Rouland, on the other hand, got his start by studying and working with his uncle, Joseph-Aignan Sigaud de la Fond.[24] The Catholic Church too, although perhaps unwittingly, acted as a patron for many scientific popularizers. Several disseminators held minor orders in the Church, usually at the level of an abbé. As many as one in ten used this title, which implied that the individual had received holy orders and held a benefice with an annual income, even though he had not been ordained in the priesthood.

Of all the popularizers who worked in the eighteenth century, the Abbé Jean-Antoine Nollet most effectively combined elite science and the spectacular, theatrical elements of popularization. At the same time, he synthesized a new pedagogy that attracted the attention and the admiration of people throughout Europe and North America. He achieved enormous popularity, far beyond that of any other popularizer, while simultaneously collecting state positions, royal patronage, and international status. Nollet, therefore, provides an excellent illustration of the possibilities for disseminators of science and the potential for creating links between elite and popular culture even though his success exceeded that of every other popular science practitioner. He served as both a prototype and a model for other popularizers seeking to imitate his success. Everybody interested in experimental physics, whether or not they agreed with his ideas, imitated Nollet's experiments and pedagogical style. Voltaire wrote that "a simple mechanic like the abbé Nollet, who does not know anything other than the new experiments, is a better physicist than Democritus and Descartes. He is not as great a man, but he knows more and better."[25] Nollet's name became a byword in France for popular science; the Cardinal François Joachim de Pierre de Bernis claimed in a letter to Voltaire that in Paris, "all women had their *bel esprit*, next their geometer, and finally their abbé Nollet."[26]

Born in the town of Pimprez on 19 November 1700 to farmers, Nollet received his education first at the Collège de Clermont in Beauvais and then at the University of Paris where he studied theology. Nollet obtained a position as a deacon in 1728, along with the title of abbé, but never attempted to further his position in the Catholic Church. Instead, Nollet sought his fortune in the realm of science. He trained himself in natural philosophy and in artisanal crafts. In particular, he studied enameling with the celebrated Jean Raux, enameler to Louis XV. He later contributed the volume on hat-making to the "Description des Arts et Métiers," a series of technical manuals on various trades. His work in the sciences landed him a place in the newly formed Société des Arts. This group, founded by Louis de Bourbon-Condé, count de Blermone, had several goals including a devotion to the application of the sciences to manufacturing arts. As noted above, he also got a job with two of the leading academicians of the time, Dufay and Réaumur. Nollet worked for them from 1731 or 1732 until 1735. During that time he traveled to England and the Netherlands where he met savants and popularizers such as John Keill, Jean-Théophile Désaguliers, and Petrus and Jan van Musschenbroek.[27]

In 1735 Nollet began to work full-time offering courses in experimental physics as well as making and selling scientific equipment. One of his early clients, Voltaire, paid Nollet about ten thousand livres for a complete *cabinet de physique* that he and Emilie du Châtelet used at her château in Cirey while they were working on the *Eléments de la philosophie de Newton*. Nollet's course earned immediate accolades as people recognized his value as a scientific educator and the facility of his teaching style. Some of his students, such as Antoine Lavoisier and Charles-Augustin Coulomb, went on to become well-known savants themselves. In addition to his courses, men and women flocked to Nollet for individual tutoring in experimental physics.[28] His success as an educator, along with the influence of his patrons, aided Nollet in gaining membership to the Académie Royale des Sciences in 1739 and, in 1746, lodgings in the Louvre.[29]

Nollet's reputation spread quickly and, in 1739, Charles-Emmanuel III, the king of Sardinia, invited him to come and present his course before the royal court. Nollet ultimately spent six months in Turin. The enormously satisfied king purchased an entire set of scientific instruments from Nollet so that both he and his son could use them. Nollet also apparently had fruitful conversations with Laura Bassi and Maria Angela Ardinghelli, two of the leading female savants in Italy. In 1744 Nollet received an invitation to give some lessons to the young Dauphin, Louis, and his wife, Marie-Thérèse de Bourbon. In addition, Louis XV, who sometimes suffered from attacks of ennui, found Nollet's lectures both entertaining and intellectually stimulating and so rewarded him with a gift of one thousand écus each year he lectured at Versailles. This almost made up for the irritation Nollet sometimes felt at being constantly summoned to lecture before the royal family when he would have preferred conducting his own research. Nonetheless, this tacit royal approval brought him further fame and notoriety. When several provincial academies decided to establish their own *cabinets de physique*, they turned to Nollet for help. Nollet provided advice and also sold them some of the necessary equipment. The Académie de Bordeaux even managed to convince Nollet to come and offer a series of sixteen lectures in 1741. The academicians found the experiments so interesting that the following year they decided to spend 1,393 livres for the purchase of scientific instruments and machines.[30]

With fame came some controversy. During the 1740s, several Italian physicists claimed to have utilized Nollet's electrical methods to heal patients. Nollet remained unconvinced and traveled to Italy to test these theories himself. The Italians uniformly failed to live up to their claims and invented rather ingenious excuses for their failures. One blamed his inability to properly electrify a young woman on Nollet's companion, a Cardinal, although exactly why he had this negative effect is unclear.[31]

By mid-century, people viewed Nollet as the experimenter *par excellence*. He gained membership in numerous academies throughout Europe; the king made him the "Professor of Experimental Physics and Natural History to the French Royal Children;" and he joined the faculty at the Collège de Navarre as Professor of Experimental Physics, a post created expressly for Nollet by Louis XV in 1753. This last honor marks a turning point in Nollet's career and in the place of popular science in France. Louis XV stipulated that this professorship should concentrate on making experimental physics available, free of charge, to whoever wanted to learn. It is at this point, then, that popular acclaim, academic recognition, and state support coalesced in the person of Nollet. His popularity whetted the appetite for French men and women at many different social levels who wanted to gain access to experimental physics. Nollet gave his inaugural lecture at the Collège de Navarre before a crowd of five hundred onlookers in a special amphitheater built for this new class.[32] He began his opening speech in typical fashion, speaking in Latin; significantly, however, he broke with tradition and gave the bulk of his oration in French, thus emphasizing from the very beginning his interest in reaching the greatest number of people.[33]

Scientific popularization had by now become a significant part of cultural life in Paris. Nollet wrote that "the taste for physics has become nearly general" adding that popularizers now found it desirable to frame their examples "at the level everybody." Many visitors to Paris attended Nollet's lectures along with their visits to Versailles, Notre-Dame, and the salons. When the traveler Joseph Teleki came to Paris in 1760 he made a point of attending some of Nollet's lectures and concluded that he was one "of the great savants" of Europe. Bengt Ferrner, traveling from Sweden, heard him speak before a crowd of five hundred people and commented favorably on the skill with which Nollet set up and performed his experiments. A guidebook written in 1763 listed both Nollet's courses at the Collège de Navarre and his private tutorials as events of particular interest to people visiting Paris. One observer at these demonstrations thought that they were "singular, amazing, [and] amusing," adding that Nollet "excited the desire to know and to understand physical causes."[34] Nollet's fame did not diminish during his lifetime. In 1769, when his book *L'Art des experiences* was printed, an anonymous reviewer wrote "the name of the author alone makes this book commendable." In that same year, Elie Fréron claimed that Nollet's work "merits the attention & the approbation of all the physicists & all the amateurs of that agreeable science." Following his death, his name became a touchstone for good experimental science for many years to come.[35]

Popularizers at work

Popularizers put forward a broad spectrum of ideas within their courses. Some disseminators dealt strictly with Newtonianism or Cartesianism. Others rejected overarching theories in favor of a focus on experiments and topics most likely to draw an audience. No matter what ideas or theories the disseminator may have wished to get across, audiences demanded that popularizers at least attempt to draw connections between their experiments and the overall utility. Popularizers designed their courses to inculcate auditors with at least an appreciation, if not a deep understanding, of natural philosophy. To achieve this end, the proprietors of one-day-only courses offered a loose conglomeration of experiments guided by the level of enjoyment they might bring to the audience. In other words, the exact order of the presentation focused on the spectacular rather than on following any specific scientific program. For most of these popularizers, little information exists regarding the specific nature of their experiments. Instead, their advertisements tended to list the possible demonstrations that they performed on any given night. One such popularizer, however, published a volume of his experiments; this is Giuseppe Pinetti's *Amusemens Physiques*. This book provides a description of forty-three experiments performed by Pinetti in his physics show. No obvious order emerges from the collection and, in fact, some of the "experiments" seem somewhat out of place since they apparently have little to do with physics. In fact, many of Pinetti's experiments had overtly theatrical names such as "the philosophical mushroom" or the "magic table." He mixed these experiments with procedures for changing the color of a rose, mathematical calculations with cards, and a method of conjuring an egg under a goblet. One experiment, called a "Card Trick," simply describes a method of counting cards in order to guess which three cards an audience member selected out of a deck. Another experiment explains how to draw a disfigured form that, from the right point of view, appears perfectly proportioned. Other experiments rely on knowledge of chemicals. Most of the experiments appear to us as little more than parlor tricks.[36]

While the one-day-only courses seemed to lack structure, the popularizers who offered courses lasting from one to four weeks did, generally, imbue their courses with a greater sense of purpose. It helped that these courses focused on a limited number of topics. Jacques Bianchi, for example, typically offered a four-week course, meeting once each week. He dealt with one topic each week. His course in August 1785, for example, covered light, color, electricity, and gases (particularly hydrogen so that he could discuss balloons). In 1782 Bianchi offered a similar course devoted entirely to electricity. He displayed the properties of electricity, using several machines of his own design, and demonstrated the connections between electricity and hydraulics, hydrostatics, optics, and chemistry. Like the one-day-only popularizers, Bianchi emphasized the spectacular, largely because of his interest in selling his machines. In addition to working as a popularizer, Bianchi also sold scientific equipment including a perpetually popular item he called a *briquet physique*, essentially a jar filled with phosphorus that gave off light.[37] However, by focusing on a more limited range of topics, it seems possible that an individual following one of Bianchi's courses would leave with a somewhat deeper sense of experimental physics than those attending Pinetti's amusing physics shows.

The longer courses in experimental physics covered the broadest range of topics. Jean-Antoine Nollet's course, which lasted the entire academic year, offered the deepest look at the topic. He also published his lectures, in six volumes, to great acclaim; his books went through numerous editions during his lifetime. Most of the popularizers at this level, however, did not offer such a long course, opting instead for courses lasting between thirty and sixty lessons, or ten to twenty weeks. Jacques-Alexandre-César Charles, for example, developed a course that met sixty times. After an introductory lecture Charles spent the next twelve lectures examining mechanics. He then gave four lectures on hydrostatics; eight on pneumatics; nine on chemistry; twelve on electricity; eight on optics; four on acoustics; and one each on balloons and magnets. Charles performed these lectures using a vast array of equipment; he possessed the finest *cabinet de physique* in Europe at the time.[38]

Collectively, popularizers offered their auditors a wide range of courses from which to choose. The structure of the classes differed somewhat depending on location and affected the price and the nature of the audience, topics I will analyze in Chapter 3. However, some patterns emerge that reveal popularizers generally followed similar formats for their work. The time of day one might attend a course could vary enormously. Popularizers experimented with different start times throughout the eighteenth century with a few being offered as early as 9:00 a.m. or 10:00 a.m. and some beginning as late as 8:00 p.m. or 9:00 p.m. Overall, however, disseminators tended to follow the patterns of the lives led by their auditors. For those potential students with leisure time on their hands, courses started between 11:00 a.m. and noon. Mathurin-Jacques Brisson, for example, offered his course at 11:00 a.m. from its inception in 1764 until the early 1780s when he experimented with beginning at 11:30 a.m. and 12:00 p.m. For those people interested in attending courses, but forced to work during the day, the second most popular start time was 6:00 p.m. While longer courses could be found both at daytime and evening hours, shorter courses, especially one-day-only courses, mostly appeared in the evenings, fitting both the time slot and the economic means of the likely auditors. Some disseminators offered the same course twice, once in the morning and again in the evening. Competition between popularizers existed. Frequently a popularizer would start a course within one hour of a rival's, on the same days and for approximately the same length of time. Joseph-Aignan Sigaud de la Fond, for example, started his courses on the same days and even at the same times as Brisson.[39]

The length of the class sessions remained relatively static. Courses were generally one-and-a-half to two hours in length with a few courses running just one hour and a few lasting longer. In some cases, such as at the annual fairs, the courses operated like open houses, with a continual run of experiments in progress that an individual could pay to observe but leave at any time. In other cases, a popularizer might have a show in which he demonstrated over a three-to-four-hour period how all of his equipment and machines operated. Again, auditors would pay to enter the *cabinet de physique*, but could probably leave in the middle if they wished.

The frequency of courses also followed distinctive patterns. One-day-only courses might be offered on any day. Some popularizers might close down for one day a week, often Sunday, but others did not. Typically, disseminators offered such courses continuously for a period of time, perhaps one or two weeks, and then closed down for a short period before resuming. Longer courses had more variety. These courses can be divided into two categories; courses that lasted one month or less and those that were longer than one month. The longer courses might last anywhere from six to twelve weeks or more. For both shorter and longer courses, the classes themselves typically met two or, most often, three days each week. Thus, a course would meet Monday, Wednesday, and Friday or Tuesday, Thursday, and Saturday but most likely not on Sundays.

Popularizers gave courses throughout the calendar year although summer courses were more rare and were of the shorter or one-day-only variety. In the eighteenth century, as now, Parisians left the city during the summer if they could. This reduced the size of the potential audience somewhat for the longer, more expensive classes. In general, courses met most frequently between November and May with most courses starting in December. During this peak period many popularizers offered the same course back to back. In 1782, Charles, for example, offered his course in experimental physics from December through February and then again from March through May. In addition, he offered the same course during the day and in the evening. Charles supplemented these longer courses, occasionally, with shorter versions devoted to specific topics. In 1786, Charles taught short courses specifically on electricity or optics that began in May or September but lasted only a few weeks.[40]

The different opportunities available for auditors reflected the motives of the popularizers and the goals of the courses. In particular, the simultaneous rise both in opportunities to teach and in public interest in experimental physics saw a new emphasis on experimentation over theory and on applied science combined with spectacle. As one contemporary noted, the success of popular science "depended not only on the authority acquired by the teacher in the sciences and his expository talents, but also on the richness of his cabinet that he had created for his experiments." The need to combine erudition with excitement led to numerous appropriations by the popularizers from both elite savants and street performers.[41]

Although scientific popularizers usually did not attach themselves to any one particular theoretical system, several did just that. As one might expect in the age of the Enlightenment, Newtonianism received some attention. In the 1760s, Allard offered a course in experimental physics that purported to explain all the principles of Newton's natural philosophy. Allard performed, so he said, a great number of "interesting and curious" experiments designed to confirm Newton's theories. Antoine Deparcieux also concentrated his attention on Newtonian astronomy and optics for the physics course he established during the 1780s. In addition, he applied Newtonianism to his explanations of other phenomena such as electricity and magnetism. In 1783 Jean-François Pilâtre de Rozier appropriated Newton's theories and "most entertaining experiments" for a series of lectures on optics. In the lectures he offered at the Musée de Monsieur he frequently expounded on Newtonian attraction, although it remains unclear just how closely he followed Newton's ideas.[42] Even when popularizers claimed to be focusing on Newtonianism, we should not jump to conclusions about the precise content of their teachings.

For every popular lecture that focused on Newton, many more proposed other theories. Allard, in addition to his explication of Newtonianism, also covered the work of such ardent Cartesians as Jean-Baptiste Dortous de

Mairan. Another lecturer, named Delor, followed the teachings of Nollet for the courses he offered during the 1760s. The abbé de la Pouyade proselytized the physical system created by Jacques-Charles-François de la Perrière de Roiffé. He rejected Newtonian gravity in favor of a theory of impulsion whereby astronomical bodies are moved by an outside force, rather than through a mysterious, as he saw it, inner attraction.[43]

Some popularizers favored their own theories over those of any of the savants who came before them. Jean Paul Marat thought that his work in such areas as optics had surpassed Newton's and "changed the face" of science. As a result he offered several courses advocating his own theories over those of other savants. Charles Rabiqueau published his ideas on electricity in a book, *Le Spectacle du feu élémentaire, ou cours d'électricité expérimentale*, in which he related more than sixty of the experiments he routinely performed for his students. Essentially, Rabiqueau argued that electricity was really just a special form of fire and that it was this fire that served as the basic mechanism of the universe. This placed him outside contemporary speculation on the nature of electricity, which focused largely on the connections between lightning and electric fluids. Nonetheless, *Le Spectacle du feu élémentaire* remained a popular work; publishers reprinted it, and journalists favorably reviewed it as late as 1785.[44] Optics provided much of the emphasis in Rabiqueau's later work. He transformed his *cabinet de physique* into a "school of vision" where he exhibited the advantages of microscopes in a series of twenty-two experiments. Just as with the earlier work on electricity, however, his optical efforts ran contrary to existing scientific thought and did not receive the kind of favorable attention as his earlier work. Rabiqueau, for his part, lamented that the Académie Royale des Sciences "refused to accept knowledge" when it was offered to them.[45]

Although some popularizers were anxious to spread their own ideas, more often they presented an eclectic blend of various theories, if they presented any theories at all. Nollet, for example, claimed that he did not limit himself to any particular theory of physics. "It is not the physics of Descartes, nor of Newton, nor of Leibniz, that I propose particularly to follow; but without any personal preference, and without any distinction, it is that which, by the general vogue, and by well attested facts, shall appear to me to be best established."[46] This does not mean that Nollet did not hold any system to be true or that he did not favor one system over another. But, in fact, there are strong elements of both Cartesianism and Newtonianism in his work.[47] Simply put, Nollet's theories tended to favor the Cartesian worldview while his emphasis on experimentation tended to rely on the Enlightenment appropriation of Newtonian methodology. Significantly, Nollet almost always emphasized the useful side of natural philosophy and focused on that part of experimental physics most accessible and available to the largest possible audience. For this reason, Nollet's views on a variety of scientific topics remained popular long after the "correct" answers had been discovered.[48]

If specific theoretical systems did not especially concern scientific disseminators, then what goals did they hope to achieve? One answer to this question comes from Nollet. In his inaugural lecture at the Collège de Navarre, Nollet discussed two main topics. First, he outlined what it meant to study experimental physics. Second, he commented on what the practitioner of this subject needed to do to succeed. "The object of experimental physics," claimed Nollet, "is to understand the phenomena of nature, and to show the causes of the phenomena through factual proofs," by which he meant experiments.[49] Nollet felt that his audience could easily reach this goal no matter what their reasons for attending his classes. Nollet had a realistic attitude towards his students; "we dare not pretend," he wrote, "we cannot even desire, that all our auditors become physicists." Nonetheless, Nollet also hoped that "among the vast majority, we will find many in whom we will create the desire to carry further that study, and who will give themselves entirely to continuing it, or at least in the moments of leisure that their professions and their affairs may leave them."[50] Nollet always endeavored to make his lessons both instructive and easy to follow and refused to weigh his lectures down with debates over the validity of any one theoretical approach. Instead, he noted that Descartes, Newton, and Leibniz, among others, were all great men but that none of them had proved infallible.[51] Nollet understood the power of experiments in persuasion and preferred to rely on observation and experimentation for his knowledge. By following his example, he claimed that anyone could become a good amateur physicist.

Demonstrations dominated most experimental physics courses while complex theories and mathematical proofs played a minimal role. The students accepted the authority of the teachers to create experiments that accurately illustrated good scientific principles. Some of the most entertaining displays came in the field of electricity; the best-known and most oft-mimicked experiments dealt with the Leyden jar and its various powers. In the early 1740s the lawyer and amateur physicist Andreas Cunaeus invented the Leyden jar, so-called after the location of its discovery, and immediately reported his findings to the Dutch physicist Petrus Musschenbroek. A glass jar or globe filled with water, it stored a substantial amount of electricity released when a rod, extending from the water in the jar, touched a conducting material. Both Cunaeus and Musschenbroek used themselves as conductors, giving themselves an electric shock. This was a painful surprise since the charge provided by the jar was much greater than had been achieved before. Musschenbroek "was so painfully affected," wrote his assistant Allamand, "that when he came to my house a few hours later he was still shaken, and he told me that nothing in the world could make him try the experiment again."[52] As noted in the Introduction to this book, Nollet famously transmitted an electrical shock through 180 of Louis XV's royal guards and, on another occasion, to 200 Carthusian monks. Nicolas-Philippe Ledru put an interesting twist to this demonstration. The volunteer subjects of his experiment came from his audience, a technique that really brought the power of electricity home to the students.[53]

Another experiment made popular by Nollet and others required the assistance of a small boy. He hung the boy from the ceiling by silk cords and electrified him by means of an electrical machine; this caused the body of the boy to act as a magnet. Various objects placed within his reach would be drawn towards the boy's outstretched hand. Frequently, he encouraged a young girl to touch the electrified boy or, better yet, to give him a kiss. He dimmed the lights of the room and the boy and the girl, facing each other, would move close enough for sparks to pass from one to the other, a phenomenon seen easily in the darkened room. This "electric kiss" was quite popular, especially, contemporaries claimed, among the ladies.[54] Electrical experiments like these, and many more besides, represent the kinds of public displays of science commonly found in the eighteenth century. Their significance lay not so much in proving certain things about electricity as in visually demonstrating science in action. These displays were easily viewed by a public unaware of the nuances behind complicated scientific instruments. At the same time, broad generalizations could be drawn from these demonstrations. In the case of the Leyden jar experiments, both the speed and strength of the electric shocks could be gauged and visually measured to some extent. In other words, these experiments promoted discussion and drew attention to important issues.

Disseminators often chose experiments that had an inherent entertainment value in their effort to combine science with spectacle. The passion for utility was indelibly connected to the theatrical side of natural philosophy. Science as spectacle had a long tradition back to the work of Rohault and Polinière, but in the eighteenth century it took on a new importance. By the 1750s, science had gone beyond its original aim of providing a fun way to perform experiments and entertain the audience. Now the spectacular nature of the demonstrations became a key selling point in promoting popular science courses. Ledru adopted a stage name for his classes; he was known as Comus, named after the Roman god of revelry. This, along with his theatrical style, earned him a reputation as something of a scientific magician. This could go too far in the eyes of some people and, indeed, the philosophe Claude-Adrien Helvétius wrote that parts of Ledru's show were "more curious than useful."[55]

Popularizers, as a general rule, provided visually appealing and educational experiments for their students with little emphasis on proving any particular theories. The work of the experimental physicist Perrin, cited in the

Introduction, provides a case in point. He performed lectures on natural philosophy to wide acclaim even though his experiments did not promote any theory at all. As he wrote in one of his advertisements, quoting an unnamed classical source, "not by words, but by deeds are the arts proven." His popularity as a disseminator of science rested squarely on the abilities of his dog. Perrin and his savant dog enjoyed an enormous success in Paris, causing him to write, "O happy century, where Science/Is an amusement for everyone!/Charmed by our intelligence/Physics has become a very enjoyable subject." Perrin had every right to be happy with physics and his success as a popularizer, but the kind of science demonstrated by this self-proclaimed "mechanic, engineer, and demonstrator of amusing physics" remains somewhat obscure. His experiments sound suspiciously like magic tricks or visual conundrums: the "enchanted tower," the "incomprehensible inkwell," the "dancing rings," and the "sympathetic light."[56] Perrin interspersed demonstrations of automata and his talented dog in between more straightforward experiments. He varied each show so that different topics appeared each night, probably to encourage return visits.

By all accounts, "entertaining physics" achieved a significant, although sometimes ambiguous, status during the eighteenth century. Rousseau, somewhat derisively, even described Nollet's *cabinet de physique* as "a laboratory of magic" in which one saw a "collection of miracles." Pilâtre de Rozier said that his optical experiments would be "amusing and illuminating." Noël, who offered bi-weekly courses in experimental physics, said that he designed his class for those looking for "entertainment" and that all of his experiments would be both "recreational and enjoyable." Although Rouland balked at describing his class as entertainment, he did state that his experiments would reveal "a number of curious and interesting applications." François Pelletier had fewer qualms than Rouland; he actually described the lectures he performed as "a spectacle" and claimed to cover many of the different "entertaining kinds" of science.[57]

Not all disseminators stopped at describing their work as entertaining. Instead of emphasizing their knowledge, these popularizers concentrated on their abilities to amuse the audience. Science could be fun only if it fell into the hands of a scientific showman. Perrin began to call himself a "teacher and demonstrator of entertaining physics." Pinetti, appearing in Paris periodically from 1783 to 1787, often performed his experimental physics show at the royal Théâtre des Menus-Plaisirs. Contemporaries apparently appreciated his efforts; one described Pinetti as "a conjuror infinitely superior to Comus." Pinetti's theatrical style effectively combined science and spectacle. Thus, while Pinetti's admirers considered him a professor of mathematics and physics, they also referred to him as a conjuror. Pierre-Jean-Baptiste Nougaret, in his guidebook to Paris, also muddled the issue of science and magic. In his description of Pinetti's activities Nougaret left the issue up to his readers by describing him as either a "very skillful physicist *or* a conjuror."[58]

In utilizing a theater for his combination of science and spectacle, Pinetti followed in the footsteps of other popularizers. Many popularizers performed scientific demonstrations and offered courses in theaters, particularly those found on the Boulevard du Temple. In 1783, Nicolet's theater was transformed, on certain days, into a "school for physicists."[59] The general popularity of these theaters probably aided in attracting an audience for the popular science shows produced there. Perrin and his dog, for example, frequently worked out of the boulevard theaters, often for several days in a row, presumably while the actors rehearsed a new play.[60]

Pinetti's theatrical and entertaining experiments seem a long way from some of the popularizers who strove to disseminate the theories of Newton or Descartes. Nonetheless, most of the popularizers working in France during the eighteenth century employed similar techniques whether they aimed at disseminating Newtonianism, exhibiting the usefulness of science, or trying to entertain. Popularizers concentrated on pleasing their audience so that they could continue to attract students and survive as purveyors of science. This was a necessary tactic since they needed the support of the general public in order to have financial independence. There were certainly not enough academic or state positions for every savant to pursue scientific research full-time. Thus, the practice of popular science provided a viable alternative to those people unable to gain patronage from traditional sources. In essence, popularizers garnered patronage from the public.

Even as popularizers combined science with spectacle they also emphasized the enlightened notion of utility. "The science of physics," wrote one commentator, "while useful and necessary in all its parts, also reaches out to amuse us and provide us . . . with some of the most interesting recreations."[61] This combination of entertainment and education occurs frequently at nearly every level of popularization; disseminators did not want to water down the knowledge to make it easier to understand but to make the learning process fun. For experimental physics, this did mean, however, that the science disseminated generally appeared without mathematics. Experiments substituted for mathematical proofs. In effect, popularizers omitted the technical side of their subject by revealing to their audiences the effects of various phenomena along with the grand ideas behind those experiments while at the same time skipping most of the work that had occurred in between.

This utility is clearly a part of the Enlightenment project, especially as it relates to the work of the *Encyclopédie*. The encyclopedists aimed to make available to as many men and women as possible the sum total of all knowledge.[62]

They deemed much of that information useful; trade and industry, for example, received a place of honor and appeared accompanied by numerous plates depicting in detail the work being described. Public lecture courses were an oral version of the *Encyclopédie*. From this perspective, disseminators of natural philosophy implied all of the aforementioned aspects of utility. Broadly speaking, the public and the state benefited from an accurate understanding of scientific principles and the inner workings of nature. More specifically, students who had attended popular science classes might apply their new knowledge to their daily lives and adapt to the natural order in an efficient manner. Disseminators often described experimental equipment, machines, and demonstrations as particularly useful or of having a special utility. Pagny, for example, suggested that his lessons were "useful for the good of society" and that his experiments both had "utility" and were "agreeable." Rouland claimed society was indebted to experimental physics because of all the "new, interesting, and precious discoveries" it had provided to humanity.[63]

The individuals who attended these classes obtained their knowledge through experiments but without mathematical justifications. Experiments demonstrated the advantages of natural philosophy, advantages both economic and intellectual. Lectures on magnets, for example, might apply to navigation, aid lost travelers roaming about France, and more generally help people understand how the world operated. This was not the same kind of economic outcome for popular science as seen in other countries, especially in England where, as Larry Stewart has shown, some popularizers of science applied their efforts towards solving specific technological problems that aided and abetted the burgeoning changes in industry. In general, both philosophes and the public in the eighteenth century considered only useful knowledge worthwhile. Diderot noted, albeit with some disdain, that the common people [*le vulgaire*] always asked to what use philosophy could be put.[64]

The notion of utility had an important place in the eighteenth century, particularly in its association with reason, science, and the encyclopedic side of the Enlightenment. The utility of popular science in France centered on its role in providing cultural capital. Natural philosophy, in particular, was useful because it could be used for social reform and the overall rationalization of society rather than because of its more concrete applications. The discourse of utility frequently found itself associated with the application of science to society and the overall development of civil society. In other forums, people linked science to notions of profit, commerce, and free trade as well as general happiness.[65]

We should not conflate, however, the rhetoric of utility with the practical advantages that popular lecture courses might give their students. Popularizers wanted their classes to attend to the needs of their students in specific and concrete ways. With this in mind, they often prepared unique experiments that would lend themselves to precise applications. On the other hand, they did not expect all individuals to go out and replicate experiments in their homes to solve day-to-day problems. Instead, popularizers hoped that their demonstrations would help certain groups of people, such as artisans and engineers, to look at their world in a different manner. At the same time, for the bulk of the individuals taking classes for their own edification, this interest in the usefulness of experimental physics might still apply. The audience would be able to learn how nature could be controlled through experiments. In an age stunned by an expanding universe and overwhelmed with invisible fluids, people eagerly accepted this demonstration of control over the natural world. However, much of the utility was still of a rhetorical nature. Popular science could be useful for the students' well-being, and would enable them to converse with other enlightened individuals or read enlightened materials, but it did not have many, if any, specific tangible benefits.

Disseminators marketed popular science, therefore, on the basis of its theoretical importance, entertainment value, and utility. This perceived utility played a strong role in the diffusion of experimental physics as it did for many other fields. At the same time certain popularizers focused on self-improvement and the "cultivation of the mind." Nollet assumed that the people who followed his lectures would be either those who were "driven by the tastes of others," or amateurs "deprived of the necessary means for abandoning themselves over to new research." In both cases, the results matched; interested people enriched their understanding of the world by attending public lectures. Mary Douglas and Baron Isherwood have argued that consumers express a "concern for information about the changing cultural scene."[66] With this in mind, it is difficult, and perhaps impossible, to know for certain if public lecture courses gained acceptance due to a change of fashion or because of their perceived utility and value. Certainly, the usefulness and practicality of experimental physics added to its reputation. Popular science combined both entertainment and utility. The article on commerce in the *Encyclopédie* emphasized this idea; its author claimed "men need instruction and amusement" in the sciences and the arts in order for the people and the state to profit. Utility remained a prevalent theme throughout the eighteenth century, especially among members of the middling classes.[67]

People greatly appreciated those popularizers who focused on the practicality of their demonstrations. Nollet's name, for example, became synonymous with good experimental physics largely because he always worked with "public usefulness" in mind. Superstition, he wrote, would diminish as "more people understand the physical causes of the effects of nature."[68] Contemporaries dubbed Nollet the "Prince of Science" and said that he symbolized all its positive aspects. As one poet wrote "Come inspire me, Doctor of Physics,/I paint the electric marvel/And its wonders that are praised./Gods, what a magnificent scene/Prepared by the hand of the Nollets!" People quickly claimed that Nollet's teaching style was perfect for the task at hand. Noël-Antoine Pluche even suggested that all experimental science courses should be modeled after Nollet's.[69] Like Nollet, other disseminators also insisted on the practical value of their experiments and classes. Pilâtre de Rozier thought that his courses "principally demonstrated the usefulness of [science]" to the mechanical arts. François Pelletier believed that his lectures advantageously displayed the pragmatic applications of various scientific instruments and machines for artisans and craftsmen. Pagny claimed his lectures on experimental physics were "useful for the well being of society" in general.[70] The utility and practicality of experimental physics occupied a central place in the efforts of many popularizers.

Popularizers operated in a complex arena where they forged connections between elite science and charlatanism, utility and frivolity, instruction and entertainment. In forging a new occupation where none had previously existed, these individuals created a new kind of scientific dissemination, one that consciously combined entertainment and instruction. They also developed connections between their work and the efforts of the proponents of the Enlightenment by explicitly suggesting that popular science courses had utility. In this way, disseminators of experimental physics carved out a place for themselves within the spectrum of scientific activity in France. In part, they succeeded thanks to their exciting pedagogy and the emergence of an audience eager to obtain what the popularizers had to sell. At the same time, however, they received support and patronage from people willing to back a group of savants working outside the select group of state-sponsored individuals. The perceived utility of science ensured that even producers and disseminators of scientific knowledge who worked outside the state system would be valued.

Notes

1 Polycarpe Poncelet, *Principes généraux pour servir à l'éducation des enfans, particulièrement de la noblesse françoise*, 3 vols. (Paris: P.G. Le Mercier, 1763), II: 265–6.

2 Pierre Polinière, *Expériences de physique*, 5th edn, 2 vols. (Paris: Clousier, 1741), I:254–6. Blake T. Hanna describes this experiment; "Polinière and the Teaching of Experimental Physics in Paris: 1700–1730," in *Eighteenth Century Studies Presented to Arthur M. Wilson*, ed. Peter Gay (Hanover: University Press of New England, 1972), 24.

3 On the Montmor Academy, in operation from 1657 to 1664, see Harcourt Brown, *Scientific Organizations in Seventeenth-Century France (1620–1680)* (New York: Russell & Russell, 1934), chps. 4–6; on Renaudot, who held his meetings from 1633 to 1642, see Kathleen Wellman, *Making Science Social: The Conferences of Théophraste Renaudot, 1633–1642* (Norman: University of Oklahoma Press, 2003); and Simone Mazauric, "La Diffusion du savoir en dehors des circuits savants: Le Bureau d'adresse de Théophraste Renaudot," in *Commercium litterarium, 1600–1750: La communication dans la république des lettres*, eds. Hans Bots and Françoise Waquet (Amsterdam: APA-Holland University Press, 1994), 151–72.

4 Aspects of this trend for high society in the period from 1650–1750 have been studied by Geoffrey Sutton, *Science for a Polite Society: Gender, Culture, and the Demonstration of Enlightenment* (Boulder, CO: Westview Press, 1995).

5 Marie Jean Antoine Nicolas de Caritat, Marquis de Condorcet, "Eloge de Rohaut [sic]," in *Oeuvres*, eds. A. Condorcet O'Conner and M.F. Arago (Paris: Firmin Didot Frères, 1847), II:95. On Rohault see Monette Martinet, "Jacques Rohault (c.1617–1672)," in *Quelques savants et amateurs de science au XVIIe siècle: Sept notices biobibliographiques caractéristiques*, eds. Pierre Costabel and Monette Martinet (Paris: Société Française d'Histoire des Sciences et des Techniques, 1986), 89–132; and Pierre Clair, *Jacques Rohault (1618–1672): Bio-bibliographie* (Paris: Editions du Centre National de la Recherche Scientifique, 1978). One historian has suggested that Clerselier forced his daughter into the marriage with Rohault, whom she saw as a social inferior, "only for the consideration of Descartes's philosophy." See Pacaut, "Le Physicien, Jacques Rohault (1620–1672)," *Mémoires de l'Académie des sciences, des lettres et des arts d'Amiens* 8 (1881), 5.

6 Alexandre Saverien, *Histoire des philosophes modernes*, 8 vols. (Paris: Bleuet and Guillaume, 1773), VI:lv.

7 On Régis see Paul Mouy, *Le Développement de la physique cartésienne, 1646–1712* (Paris: Vrin, 1934), 145–67; and Richard A. Watson, *The Downfall of Cartesianism, 1673–1712* (The Hague: Martinus Nijhoff, 1966).

8 On Rohault's Wednesdays see Mouy, *Le Développement de la physique cartésienne*, 108–13. On his scientific views see Trevor McClaughlin, "Le Concept de science chez Jacques Rohault," *Revue d'histoire des sciences* 30 (1977): 225–40. On Huygens see Huygens, *Le Séjour de Christiaan Huygens à Paris*, ed. Henri L. Brugmans (Paris: Droz, 1935), 128–61. For the people in attendance see Claude Clerselier, ed., *Lettres de M. Descartes* (1659; Paris, 1724), vol. 2, preface, cited in Londa Schiebinger, *The Mind Has No Sex? Women in the Origins of Modern Science* (Cambridge, MA: Harvard University Press, 1989), 23. On the reception of Rohault's work see John Heilbron, *Electricity in the Seventeenth and Eighteenth Centuries: A Study of Early Modern Physics* (Berkeley: University of California Press, 1979), 39, 158.

9 Sutton, *Science for a Polite Society*, 114. Rohault also earned considerable fame for his textbook, the *Traité de physique*, one of the most important works of popular science ever written: Rohault, *Traité de physique*, 2 vols. (Paris, 1671). The *Traité* went through no fewer than twelve editions in France alone and was translated into both Latin and English. In addition, Rohault's treatise was cited more than any single work by Descartes or Gassendi during the seventeenth century. See Henri-Jean Martin, *Livre, pouvoirs et société à Paris au XVIIe siècle (1598–1701)* (Geneva: Droz, 1969), 933. On the publishing history see Michael A. Hoskin, "'Mining all Within': Clarke's Notes to Rohault's *Traité de physique*," *The Thomist* 24 (1961): 353–63. On the place of this work in the history of scientific textbooks see George Sarton, "The Study of Early Scientific Textbooks," *Isis* 38 (1947): 137–48; and Geert Vanpaemel, "Rohault's *Traité de physique* and the Teaching of Cartesian Physics," *Janus* 71 (1984): 31–40.

10 On Polinière see Hanna, "Polinière and the Teaching of Experimental Physics," 13–39. On his contributions to education see L.W.B. Brockliss, "Aristotle, Descartes and the New Science: Natural Philosophy at the University of Paris, 1600–1740," *Annals of Science* 38 (1981), 65; and Brockliss, *French Higher Education in the Seventeenth and Eighteenth Centuries: A Cultural History* (Oxford: Clarendon, 1987), 189.

11 See David W. Corson, "Pierre Polinière, Francis Hauksbee, and Electroluminescence: A Case of Simultaneous Discovery," *Isis* 59 (1968): 402–13; and Gad Freudenthal, "Early Electricity Between Chemistry and Physics: The Simultaneous Itineraries of Francis Hauksbee, Samuel Weil, and Pierre Polinière," *Historical Studies in the Physical Sciences* 11 (1981): 203–29.

12 The quote is from the "Abregé de la vie de M. Polinière," an anonymous appendix found in Polinière, *Expériences de physique*, 5th edn, 2 vols. (Paris: Clousier, 1741), II:n.p. For more on Polinière's course and subsequent publication of his lectures see Gad Freudenthal,

"Littérature et sciences de la nature en France au début du XVIIIe siècle," *Revue de Synthèse* 100 (1980): 267–95.

13 On the popularity of the book see the generally favorable reviews and notices it received, e.g., *Journal de Trévoux* 10 (1710): 1154–64; and *Affiches de province*, 1 May 1771, 71.

14 Public courses were advertised in a number of different newspapers, most importantly the *Affiches de Paris*, the *Affiches de province*, the *Avant-Coureur*, and the *Journal de Paris*. For a list of all the periodicals consulted for this study see the bibliography. On the value of the periodical press as historical documents see Jack Censor, *The French Press in the Age of Enlightenment* (London: Routledge, 1994); Jeremy Popkin, *Revolutionary News: The Press in France, 1789–1799* (Durham: Duke University Press, 1990); and Jean Sgard, ed., *La Presse provinciale au XVIIIe siècle* (Grenoble: Centre de Recherches sur les Sensibilités, 1983). For the affiches as a source for studying the attitudes of the middling sorts see Colin Jones, "The Great Chain of Buying: Medical Advertisement, the Bourgeois Public Sphere and the Origins of the French Revolution," *American Historical Review* 101 (1996): 13–40. On advertisements see Christopher Todd, "French Advertising in the Eighteenth Century," *Studies on Voltaire and the Eighteenth Century* 266 (1989): 513–47.

15 On the popularity of natural history see Yves Laissus, "Le Jardin du Roi," and "Les cabinets d'histoire naturelle," in *Enseignement et diffusion des sciences en France au XVIIIe siècle* (Paris: Hermann, 1964), 287–342 and 659–712; E.C. Spary, *Utopia's Garden: French Natural History from Old Regime to Revolution* (Chicago: University of Chicago Press, 2000); and Louise E. Robbins, *Elephant Slaves and Pampered Parrots: Exotic Animals in Eighteenth-Century Paris* (Baltimore: Johns Hopkins University Press, 2002).

16 On the teaching of science in Jesuit colleges see R.R. Palmer, *The Improvement of Humanity: Education and the French Revolution* (Princeton: Princeton University Press, 1985), 15. On the high turnover rate of professors see Brockliss, *French Higher Education*, 46–50. On the vacancies in former Jesuit colleges see Jean Torlais, *Un physicien au siècle des lumières: L'abbé Nollet, 1700–1770* (Paris: Sipuco, 1954), 226–28. On the Jesuit schools see Jean Morange and Jean-François Chassaing, *Le Mouvement de réforme de l'enseignement en France, 1760–1798* (Paris: Presses Universitaires de France, 1974); Charles R. Bailey, "French Secondary Education, 1763–1790: The Secularization of Ex-Jesuit Colleges," *Transactions of the American Philosophical Society* 68 (1978): 1–124; and Dominique Julia, "Les Professeurs, l'église et l'état après l'expulsion des Jésuites, 1762–1789," in *The Making of Frenchmen: Current Directions in the History of Education in France, 1679–1979*, eds. Donald N. Baker and Patrick J. Harrigan (Waterloo, Ontario: Historical Reflections Press, 1980), 459–81.

17 *Affiches de province*, 16 January 1771, 10–11.

18 *Avant-Coureur*, 18 November 1765, 721.

19 On patronage and the status of academicians see David J. Sturdy, *Science and Social Status: The Members of the Académie des Sciences, 1666–1750* (New York: Boydell, 1995); Roger Hahn, "Changing Patterns of Support of Scientists from Louis XIV to Napoleon," *History and Technology* 4 (1987): 401–11.

20 *Almanach Royal* (Paris, 1780), 466 (on Brisson). Brisson censored books in the fields of natural history, medicine, and chemistry. For Nollet's *cabinet de physique* see Archives Nationales (AN), F12–1219, dos. 3, f. 30.

21 "Précis sur la vie & des Travaux de M. l'Abbé Nollet," *Journal des beaux-arts et des sciences*, July 1770, 164 (on Nollet in Italy); *Journal encyclopédique*, July 1769, 271 (Christian VII); Derek Beales, *Joseph II*, 3 vols. (Cambridge: Cambridge University Press, 1987), I:378 (on Ledru); and *Affiches de province*, 13 March 1781, 43–4 (Pelletier).

22 Torlais, *Un physicien*, 51 (on Ledru); *Musée, autorisé par le gouvernement, sous la protection de Monsieur et de Madame* (N.p., n.d.), 3 (on Pilâtre de Rozier).

23 On Bienvenu see Patrice Bret, "Un bateleur de la science: Le 'machiniste-physicien' François Bienvenu et la diffusion de Franklin et Lavoisier," *Annales historiques de la Révolution française* 338:4 (2004): 95–127.

24 *Affiches de Paris*, 2 January 1781, 13 (on Rouland and Sigaud de la Fond).

25 Voltaire, *Notebooks*, ed. Theodore Besterman, in *The Complete Works of Voltaire*, eds. W.H. Barber and Ulla Kölving, vols. 81/82 (Oxford: Alden, 1968), LXXXI: 352.

26 Cardinal François Joachim de Pierre de Bernis, in Voltaire, *Correspondance*, XLIX: 139–40; quoted in Sutton, *Science for a Polite Society*, 225.

27 For biographical information about Nollet see Jean Torlais, *Un physicien*; Jean-Paul Grandjean de Fouchy, "Eloge de M. l'Abbé Nollet," *HARS* (1770): 121–36; and Victor-Lucien-Sulpice Lecot, *L'Abbé Nollet de Pimprez* (Noyon: Cottu-Harlay, 1856). On his membership in the Société des Arts see Torlais, *Un physicien*, 17–19. On his description of making hats see Michael Sonenscher, *The Hatters of Eighteenth-Century France* (Berkeley: University of California Press, 1987); and Arthur H. Cole and George B. Watts, *The Handicrafts of France as Recorded in the Description des arts et métiers, 1761–1788* (Boston: Harvard Graduate School of Business, 1952).

28 For Voltaire's purchase of equipment from Nollet see Voltaire to Bonaventure Moussinot, 18 May 1738, in Voltaire, *Correspondance*, VII:177. On Lavoisier see Maurice Crosland, *In the Shadow of Lavoisier: The "Annales de Chimie" and the Establishment of a New Science* (Oxford: The Alden Press, 1994), 43. For Coulomb see C. Stewart Gillmor, *Coulomb and the Evolution of Physics and Engineering in Eighteenth-Century France* (Princeton: Princeton University Press, 1971), 16. Angélique Diderot, the niece of Denis Diderot, took tutorials from Nollet while still in her teens: see Monique Ruffet, "La Physique pour debutants: Angélique Diderot et les leçons de l'abbé Nollet," *Recherches sur Diderot et sur l'Encyclopédie* 13 (1992): 57–78.

29 On his lodging in the Louvre see AN, O1–90, f. 80.

30 On Nollet's accumulation of patronage see David Sturdy, *Science and Social Status*, 389–91. On his lectures before the King and royal family Michel Antoine, *Louis XV* (Paris: Fayard, 1989), 479. On his occasional irritation see Nollet to Jean Jallabert, 19 July 1745, in *Théories électriques du XVIIIe siècle: Correspondance entre Nollet et Jallabert*, ed. Isaac Benguigui (Geneva: Georg, 1984), 120–2. On Nollet's visit to Italy see G. Hector Quignon, "L'Abbé Nollet, physicien: Son voyage en Piémont et en Italie (1749)," *Mémoires de l'Académie d'Amiens* 51 (1904): 473–539; and Paula Findlen, "Translating the New Science: Women and the Circulation of Knowledge in Enlightenment Italy," *Configurations* 3 (1995), 194–5. On Nollet's visit to Bordeaux see Torlais, *Un physicien au siècle des lumières*, 57–8; and Pierre Barrière, *L'Académie de Bordeaux: Centre de culture internationale au XVIIIe siècle (1712–1792)* (Bordeaux: Editions Bière, 1951), 33–5, 97–8.

31 Heilbron, *Electricity in the Seventeenth and Eighteenth Centuries*, 354.

32 For Nollet's title "Master of Physics and Natural History" see AN, O1–102, f. 256.

33 Jean-Antoine Nollet, *Oratio habita a Joanne-Antonio Nollet* (Paris: Regis, 1753).

34 Nollet, *Leçons de physique expérimentale*, 6 vols. (Leipzig: Arkstee & Merkus, 1754–1765), I: ix. Joseph Teleki, *La cour de Louis XV*, ed. Gabriel Tolnai (Paris: Presses Universitaires de France, 1943), 64. Bengt Ferrner, *Resa i Europa*, ed. Sten G. Lindberg (Uppsala: Almqvist & Wiksells, 1956), 352 (translation courtesy of Nancy Oblanas). Jèze, *Etat ou Tableau de la ville de Paris* (Paris: Prault, 1763), 194 (for the guidebook). *Journal encyclopédique*, July 1769, 271 (for demonstrations as singular).

35 *Affiches de Paris*, 19 December 1769, 1078 (anonymous); *L'Année littéraire*, 1769, VII: 273 (Fréron); *Avis divers: supplement de Annonces, affiches et avis divers*, 17 February 1778, 216 (reputation after his death).

36 Jean-Joseph Pinetti, *Amusemens physiques, et différentes experiences divertissantes*, 3rd edn (Paris: Gattey, 1791), 50–1 (philosophical mushroom), 68–70 (magic tables), 28–9 (cards), and 3 (figure).

37 *Journal de Paris*, 2 August 1785, 879; *Annonces, affiches et avis divers*, 25 February 1782, 450–3; François-Marie Mayeur de Saint-Paul, *Tableau du nouveau Palais Royal*, 2 vols. (London, 1788), II:17–19.

38 Two outlines of Charles's lectures are available in manuscript at the Bibliothèque de l'Institut, Ms. 2014, Pièce 1–17 (for his 1784 lectures) and Pièce 18 (for his 1802 lectures). An inventory of his *cabinet de physique* can be found in the AN, C144, f. 173.

39 On 25 February 1782, Brisson and Sigaud de la Fond both started courses that lasted for twelve weeks and met on Mondays, Wednesdays, and Fridays; the only difference was that Brisson's started at 11:30 a.m. and Sigaud de la Fond began his at 12:00 p.m. *Affiches de Paris*, 18 February 1782, 391 (Sigaud de la Fond) and *Journal de Paris*, 18 February 1782, 193 (Brisson).

40 See *Affiches de Paris*, 19 November 1781, 2671 (for December courses); *Affiches de Paris*, 4 March 1782, 512 (for March courses); *Journal de Paris*, 3 June 1783, 643 (for summer course). Thomas Bugge, *Science in France in the Revolutionary Era*, ed. Maurice Crosland (Cambridge, MA: Harvard University Press, 1969), 166–8; and *Affiches de Paris*, 18 May 1786, 1302 (for shorter courses on electricity and optics).

41 Gustave Isambert, *La vie à Paris pendant une année de la Révolution (1791–1792)* (Paris: Félix Alcan, 1896), 165 (on success of popular science); Maurice Daumas, *Les cabinets de physique au XVIIIe siècle* (Paris: Conférence du Palais de la Découverte, 1951), 21 (on the appropriations of popularizers). On the use of spectacle in scientific popularization see Simon Schaffer, "Natural Philosophy and Public Spectacle in the Eighteenth Century," *History of Science* 21 (1983): 1–43.

42 *Affiches de Paris*, 1 December 1766, 919; *Avant-coureur*, 24 November 1766, 747 (on Allard); *Affiches de Paris*, 25 November 1782, 2722 (on Deparcieux); and *Journal de Paris*, 13 April 1783, 429 (on Pilâtre de Rozier). For Pilâtre de Rozier's discussion of attraction see Antoine Tournon de la Chapelle, *La Vie et les mémoires de Pilâtre de Rozier, Ecrits par lui-même* (Paris: Belin, 1786), 11–12; and see chapter 5, this book.

43 *Avant-coureur*, 3 January 1763, 3–4 (on Allard); *Avant-coureur*, 23 November 1761, 737–8 (on Delor); *Avant-coureur*, 29 December 1766, 823–5; and *l'Année littéraire* (1766) VIII:142 (on de la Pouyade).

44 *Courier de l'Europe*, 14 March 1781, 170–1; and *Affiches de Paris*, 11 December 1784, 3280 (on Marat). Rabiqueau, *Le Spectacle du feu élémentaire, ou cours d'électricité expérimentale* (Paris: Jombert, Knapen, & Duchesne, 1753). For reviews see *l'Année littéraire* (1755) VIII:277–86 and *Journal de Paris*, 25 January 1785, 102.

45 Rabiqueau, *Le Microscope moderne* (Paris: Belin, 1785); Rabiqueau, *Description de l'école de la vision* (Paris: Chez l'Auteur, 1783); Rabiqueau, *Manifeste littéraire, servant de supplement aux Journaux sur le livre du Microscope moderne* (Paris: Chez l'Auteur, 1781), 15 (on Academy of Sciences).

46 Nollet, *Leçons de physique expérimentale*, I:xviii.

47 I.B. Cohen sees Nollet as a Cartesian; this enables him to read the Nollet/Franklin controversy over the nature of electricity as a debate between Cartesians and Newtonians. See Cohen, *Franklin and Newton* (Philadelphia: The American Philosophical Society, 1956), 387. For a different view see R.W. Home, "The Notion of Experimental Physics in Early Eighteenth-Century France," in *Change and Progress in Modern Science*, ed. Joseph C. Pitt (Dordrecht: D. Reidel, 1985), 111.

48 Nollet's views on electricity, in particular, were still being defended well into the 1780s even though Franklin's book on electricity had been available for more than thirty years at that point. See, e.g., Abbé Durand, *Le Franklinisme réfuté, ou remarques sur la théorie de l'électricité* (Paris: Varin, 1788). On this topic see R.W. Home, "Electricity in France in the Post-Franklin Era," in *Proceedings of the XIVth International Congress of the History of Science* (Tokyo: Science Council of Japan, 1975): 1–4.

49 Nollet, *Oratio*, 8.

50 Nollet, *Oratio*, 10.

51 Nollet, *Oratio*, 20.

52 See Heilbron, *Electricity in the Seventeenth and Eighteenth Centuries*, 313–14.

53 On Nollet's experiment see Joseph Priestley, *The History and Present State of Electricity*, 2 vols, ed. Robert E. Schofield (1775; rpt. New York: Johnson Reprint Company, 1966), I:124–9. *Journal de Paris*, 4 May 1780, 515 (on Ledru).

54 On the electrified boy and the electric kiss see Nollet, *Essai sur l'électricité des corps*, 5th edn

(Paris: Guerin, 1746), 76–82; Heilbron, *Electricity in the Seventeenth and Eighteenth Centuries*, 265–7; Sutton, *Science for a Polite Society*, 304–5. This experiment was reproduced well into the nineteenth century; see e.g. Jean-Sébastien-Eugène Julia de Fontenelle, *Manuel de physique amusante, ou, Nouvelles recreations physiques* (Paris: Rerot, 1829), 106.

55 Claude-Adrien Helvétius, *A Treatise on Man*, trans. W. Hooper, 2 vols. (New York: Burt Franklin, 1810), I:265. On Ledru see Jean Torlais, "Un prestidigitateur célèbre, chef de service d'électrothérapie au XVIIIe siècle, Ledru, dit Comus (1731–1807)," *Histoire de la médecine* 5 (1955): 13–25.

56 Perrin, *Amusemens physiques* (Paris: P. de Lormel, 1789), n.p. (for classical quotation); *Journal de Paris*, 12 October 1784, 1208 (savant dog); Perrin, *Prospectus* (Paris: P. de Lormel, 1786) (for poem); for the different experiments see *Affiches de Paris*, 2 April 1793, 1408; and *Chronique de Paris*, 9 April 1790, 396.

57 Jean-Jacques Rousseau, "Lettres écrits de la montagne," in *Oeuvres complètes*, eds. Bernard Gagnebin and Marcel Raymond, 4 vols. (Paris: Gallimard, 1959–), III:739. *Journal de Paris*, 13 April 1783, 429 (on Pilâtre de Rozier); *Journal de Paris*, 1 June 1783, 637; and *Journal de Paris*, 10 July 1780, 783–4 (on Noël); *Affiches de province*, 2 September 1786, 419 (on Rouland); *Affiches de Bordeaux*, 8 April 1773, 64bis (on Pelletier).

58 *Affiches de Paris*, 1 April 1787, 927 (on Perrin). Louis Petit de Bachaumont, *Mémoires secrets pour servir à l'histoire de la République des Lettres en France depuis 1762 jusqu'à nos jours*, 36 vols. (London: John Adamson, 1784–1789), XXIV:90 (comparison between Comus and Pinetti). Interestingly, Pinetti acknowledges the work of Ledru and his son in the third edition of his book. See Pinetti, *Amusemens physiques*, iii–iv. Nougaret, *Tableau mouvant de Paris, ou, Variétés amusantes*, 3 vols. (Paris: Duchesne, 1787), I: 218 (the emphasis is mine).

59 *Mémoires secrets*, 24 January 1783, 50. On popular theaters see Michele Root-Bernstein, *Boulevard Theater and Revolution in Eighteenth-Century Paris* (Ann Arbor, MI: UMI Research Press, 1984), 41–75; Robert Isherwood, *Farce and Fantasy: Popular Entertainment in Eighteenth-Century Paris* (Oxford: Oxford University Press, 1986), 167–97. For the location of these theaters see the map in André Tissier, *Les Spectacles à Paris pendant la Révolution* (Geneva: Droz, 1992), 51.

60 In 1785, for example, Perrin worked out of the Salle des Elèves de l'Opéra on the Boulevard du Temple every day from 31 March to 4 April. See the *Affiches de Paris*, 3 March 1785, 855; ibid., 1 April, 1785, 864, ibid., 2 April 1785, 872; ibid., 3 April 1785, 887; and ibid., 4 April 1785, 896.

61 A.T. Chevignard de la Pallue, *Idée de monde, ou idées générales des choses dont un jeune homme doit être instruit*, 2 vols. (Dijon: L.N. Frantin, 1779), I:230.

62 On the gendered nature of utility see Lisbet Koerner, "Women and Utility in Enlightenment Science," *Configurations* 3 (1995): 233–55.

63 On Pagny see *Affiches de Paris*, 7 January 1754, 14 and ibid., 3 January 1757, 6. On Rouland see *L'Année litteraire*, 1780, V:280–1.

64 On popular science in England see Larry Stewart, *The Rise of Public Science: Rhetoric, Technology, and Natural Philosophy in Newtonian Britain, 1660–1750* (Cambridge: Cambridge University Press, 1992). On the idea of utility see the entry in the *Encyclopédie; ou dictionnaire raisonné des sciences, des arts et des métiers*, eds. Denis Diderot and Jean d'Alembert, 5 vols., compact ed. (New York: Pergamon Press, 1985), III:558; and Jean-François Feraud, *Dictionnaire critique de la langue française*, 3 vols. (1787; reprint, Tübingen: M. Niemeyer, 1994), III:769. Denis Diderot, "Pensées sur l'interprétation de la nature," in *Oeuvres completes*, ed. Jean Varloot (Paris: Hermann, 1975–), IX:41 (on common people).

65 Roger Hahn, *The Anatomy of a Scientific Institution: The Paris Academy of Sciences, 1666–1803* (Berkeley: University of California Press, 1971), 68; and Robin Briggs, "The Académie Royale des Sciences and the Pursuit of Utility," *Past and Present* 131 (1991): 38–88 (on the utility of the Académie Royale des Sciences); Mary Terrall, *The Man Who Flattened the Earth: Maupertuis and the Sciences in the Enlightenment* (Chicago: University of Chicago Press, 2002), 164–8; Alfred Cobban, *In Search of Humanity: The Role of the Enlightenment in Modern History*

(London: Jonathan Cape, 1960), 126–32 (on utilitarians in France); Keith Baker, *Condorcet: From Natural Philosophy to Social Mathematics* (Chicago: University of Chicago Press, 1975), 55, 65–72, 75; and Keith Baker, *Inventing the French Revolution* (Cambridge: Cambridge University Press, 1990), 25–27, 163 (on science, social reason and civil society); and *Correspondance littéraire, philosophique et critique*, 16 vols., ed. Maurice Tourneux (Paris: Garnier, 1877–1882), XIII:113 (on links to profit, etc.).

66 Nollet, *Oratio*, 10. Douglas and Isherwood, *The World of Goods* (New York: Basic Books, 1979), 95.

67 On the topic of science and fashion see David Allen, "Tastes and Crazes," in *Cultures of Natural History*, eds. N. Jardine, J.A. Secord, E. Spary (Cambridge: Cambridge University Press, 1996), 394. *Encyclopedia: Selections*, eds. Nelly S. Hoyt and Thomas Cassirer (New York: Bobbs-Merrill, 1965), 70. Roger Chartier has noted that the libraries of merchants and other middling bourgeoisie often contained books that emphasized utility. See Chartier, *The Cultural Uses of Print in Early Modern France*, trans. Lydia G. Cochrane (Princeton: Princeton University Press, 1987), 196.

68 *Affiches de province*, 1 August 1753, 122 (on public usefulness); Nollet, "Un manuscript de l'abbé Nollet demandant la creation d'une chaire de physique expérimentale à la Bibliothèque royale," in Nollet, *Cours de physique expérimentale* (Paris: P.G. le Mercier, 1735). This is a handwritten letter appended to the back of the copy of the pamphlet found in the reserve room of the Bibliothèque national (book code: rès. M-R-199).

69 Abbé d'Angerville, "L'électricité," *Journal de Trévoux* 62 (August 1762), 2027–8 (poem); and Noël-Antoine Pluche, *Le spectacle de la nature*, 8 vols. (The Hague: J. Neaulme, 1739), IV:452. Not everyone praised Nollet. The Baron de Montesquieu, for example, wrote that Nollet was "a man without education." Montesquieu to President Barbot, in Charles-Louis Secondat, baron de Montesquieu, *Correspondance*, 2 vols. Ed. François Gebelin (Paris: Honoré Champion, 1914), I:358.

70 *Journal de Paris*, 4 March 1782, 249–50 (on Pilâtre de Rozier); *Journal de Paris*, 5 May 1785, 508 (on Pelletier); and *Affiches de Paris*, 7 January 1754, 14 (on Pagny).

3

The audience, economics, and geography of popular science

Popular science gripped the imagination of people all over Europe in the eighteenth century and individuals peppered their conversations with facts, allusions, references, and analogies to current scientific discoveries and debates. When John Adams arrived in France to assume his new post as United States ambassador he immediately met scientifically literate people. Adams, who was a bit less versed in the ways of sociability than some of his predecessors, especially Benjamin Franklin, found himself in a social situation where his knowledge of the sciences led to a rather unusual conversation. One night at dinner, while in Bordeaux making preparations for his journey to Paris, "one of the most elegant ladies at the table" brazenly asked him (because his name descended from "Adam") if he knew "how the first couple found out the art of lying together?" Adams, flustered by a question he found lacking in all modesty, delicacy, and dignity, chose to answer her question by making an analogy to science. "I rather thought it was by instinct, for there was a physical quality in us resembling the power of electricity or of the magnet, by which when a pair approached within a striking distance they flew together like the needle to the pole or like two objects in electric experiments." "Well, I knew not how it was," the lady responded, "but this I know – it is a very happy shock."[1] Clearly, electricity had entered into urban culture and natural philosophy, more generally, had become a topic of general conversation.

As Adams discovered to his extreme discomfiture, French people appropriated science into multiple aspects of their lives and utilized their understanding of new scientific topics in conversations, poems, plays, and literature. Popularizers targeted individuals from a variety of social and cultural backgrounds as the proper audience for their courses and sold science as a necessary and useful subject for people to want to acquire. In addition, Parisians developed the ability to judge qualitatively the different geographies of eighteenth-century science. From the Left Bank of Paris, for example, with its colleges and other academic establishments, science looked a little different than it did on the Boulevard du Temple where jugglers performed in the streets and entertainment ruled the day.

The audience

The enduring popularity of the professors and demonstrators who worked in France testifies to their ability to capture the admiration and loyalty of their audience. Joseph-Aignan Sigaud de la Fond's fame grew to such a point that, in 1777, a young artist named Coron engraved and sold a portrait of the famed popularizer, patterned after a similar engraving of Nollet, for twelve sous. The same held true for Rabiqueau, whose picture could be purchased for the same price at the shop of Legros, a master stationer. But while the popularity of the disseminators seems clear enough, the social makeup of their audience remains less obvious.[2] Enrollment figures or lists of audience members do not exist for popular science courses.

Some popularizers certainly claimed that they intended to reach as broad a public as possible. In the preface to his *Leçons de physique expérimentale*, Nollet noted that because of the general interest in physics, it behooved him to describe its principals "at the level of everybody." Although Nollet used this literary device in his works, he probably did not mean what he said. Later in the same volume, Nollet narrowed his targeted audience by admitting that his lessons were "principally destined to young people, of one or the other sex, who had passed their first years of life in the colleges or the boarding schools, to whom everything is new in nature, whose minds are naturally eager for this sort of knowledge. . . ."[3] Designing his classes and books for people who already had a basic education, however vague or inadequate, remained Nollet's standard approach throughout his life. Overall, however, popularizers targeted a fairly broad range of individuals as potential auditors.

Popularizers frequently mentioned students as people who may want to attend their courses. This fits with the initial location of popular science courses on the Left Bank. Nollet, Brisson, Sigaud de la Fond, and the other teachers and demonstrators who established their *cabinets de physique* in that section of Paris all looked to students as key participants in their courses. Pagny, for example, mentioned in his advertisements that he hoped to address his lectures in 1754 and 1755 to students in medicine and surgery. For their benefit, Pagny had a new amphitheater constructed that accommodated "a great number of people." However, the experiments he performed there, and especially his demonstrations of electricity, applied to both medicine and, more generally, to "the needs and the conveniences of life."[4] Among these students we can occasionally find some who went on to achieve their own fame. This early experience so impressed the chemist Antoine Lavoisier, for example, that during the French Revolution he recommended experimental physics courses for students as young as those attending primary schools. He claimed that it "furnishes for all the arts, and to all men, in whatever circumstances they find themselves, the instruments that are necessary for them [to deal with their situation]."[5]

While students were popular targets, advertisements for lecture courses also frequently cited that indefinable segment of the population called amateurs and provincials. Nollet addressed his last book, *L'Art des expériences*, to "the amateurs of physics, obliged or curious to do experiments themselves." In addition to those pupils he taught at the Ecole Royale Militaire, Allard also offered private courses initially "for the convenience of amateurs." He recognized, of course, that these amateurs had different needs. Some came to Paris from the provinces to attend courses while others wanted to buy good physics instruments and descriptions of experiments to take back to their homes and, subsequently, perform their own experiments in mechanics or optics. For his part, Sigaud de la Fond felt that his experiments "must interest the savants, by the important applications, and the amateurs, by the singular nature of their results." This revealing division indicates that the amateur audience for popular science, according to Sigaud de la Fond, expressed more interest in the displays and experiments than in the meaning behind them and their actual use. In any case, in 1772, a group of esteemed individuals (by their own account) engaged the services of Sigaud de la Fond to instruct them in experimental physics; they invited other interested amateurs to join in and round out the group. Like Nollet, Sigaud de la Fond singled out those keen amateurs as the group most likely to want assistance in the procurement of the necessary machines and instruments to perform experiments on their own. Disseminators identified amateurs as consumers of both knowledge and the instruments to replicate that knowledge. The popularizer Noël assumed that his electrical experiments appealed to amateurs, and Miollan claimed that his course in experimental physics suited them best as well. Addressing their classes to amateurs, however, appeared rather vague and most popularizers felt it worthwhile and necessary, perhaps to help their audience identify themselves, to indicate more explicitly the type of individual they hoped to address.[6]

Advertisements frequently mentioned visitors from the provinces. Bienvenu, for example, claimed to have received many letters requesting help in purchasing and using instruments and so became involved in setting up physics cabinets for provincial academies. Consequently, he made sure that his lectures demonstrated those machines capable of "ornamenting the most curious of cabinets." The authors of guidebooks encouraged travelers and visitors to Paris from the French provinces or even from beyond the borders of France to attend popular science classes. These authors regarded such courses as one of the attractions of Paris, and books detailing the sights of the capital never failed to mention public lectures.[7]

More specific social groups were also singled out as potential auditors. Royalty and members of the aristocracy, for example, commonly witnessed Nollet's demonstrations, and both he and Brisson lectured for the royal children. This trend extended well beyond the French royal family and included visiting royals such as the King of Denmark, who witnessed Nollet at work in 1768. In addition, Pagny gave a lecture on the uses of microscopes for the King of Poland and the Duke of Lorraine that earned the attention of scientific connoisseurs and the applause of the Académie Royale des Sciences. Joseph II of Austria paid a visit to Ledru's *cabinet de physique* to watch some electrical experiments. Noblemen and women could certainly attend these courses and, from all accounts, they did so quite frequently. Science and luxury were not at odds. In 1760, the duc de Berry attended courses given by Nollet for the Dauphin, the future Louis XVI. Loménie de Brienne, the Archbishop of Toulouse, had Deparcieux come out to his château to entertain his guests over a six-to-eight-week period in the autumn of 1778. At the same time, notables did not necessarily limit themselves to those lecturers housed in the more elite parts of Paris. The duc de Croy and several of his noble friends, for example, attended lectures given by Ledru on the Boulevard du Temple.[8]

Most often, however, a middling audience found itself the focus of advertisements for public lecture courses. While nobles and the elite provided early and constant support for popular science, the need to earn a living necessitated the participation of a much larger audience base. Unfortunately, however, lists are not available for these students. Instead, only a few scattered references have survived. A lawyer named Alix, for example, wrote a letter to the editors of the *Journal de Paris* in which he claimed to have taken one of Nollet's classes "with pleasure."[9]

Women also penetrated this scientific environment quite early on in the development of lecture courses and remained one of the groups most frequently cited by popularizers as potential auditors. Women, for example, flocked early on to attend lectures and private tutorials offered by Nollet. Although he only started giving lectures in the late 1730s, his popularity with women rose so high by the 1740s that some men stayed away. One observer facetiously wrote "among the fashionable, it was only permitted for women to get involved in physics publicly."[10] Emilie du Châtelet noted that Nollet had "at his door the carriages of duchesses, peers, and pretty women," adding that with him natural philosophy had come to seek its fortune in Paris. Francesco Algarotti, himself the author of a book popularizing Newtonianism for women, reported that Parisian ladies visited Nollet "to watch him refract light the way they go see the staging of Voltaire's *La Zaïre*."[11] Popularizers also targeted noblewomen. One observer claimed that many people of both sexes followed various scientific courses from the Parisian popularizers and that many upper-class women occupied their time with these courses for as many as a dozen years. Sigaud de la Fond numbered Madame de Genlis and the comtesse d'Harvillé among his clientele. The comtesse de Sabran attended three experimental physics courses in her lifetime. As a young woman, Henriette-Lucie-Dillon La Tour du Pin, along with her lady-in-waiting, attended classes from the abbé Pierre Bertholon while living in Montpellier. During these sessions, which met three times each week, La Tour du Pin assisted with the experiments and received individualized instruction. Her lady-in-waiting, on the other hand, spent this time cleaning Bertholon's instrument collection.[12]

Women and science: opportunities and boundaries

Many people in general, and women in particular, actively enrolled in popular science courses. Joseph-Aignan Sigaud de la Fond claimed that in the thirty years he taught experimental physics in Paris, "women had nearly always been the major part of [his] audience and always distinguished themselves by their assiduity."[13] This phenomenon was not new, however, and some of the impetus for targeting women for popular science stems from similar work done during the seventeenth and early eighteenth centuries. As we saw earlier, Jacques Rohault's Cartesian Wednesdays always had a significant number of women in the audience. Experimental physics did not monopolize the participation of women and many branches of the sciences found themselves inundated with female auditors. During the last quarter of the seventeenth century, women flooded into Nicolas Lémery's courses in chemistry, for example, causing one anxious observer to bemoan the fact that women, "trained by fashion, had the audacity to come and appear among such scientific groups."[14] The participation of women also stretched to include their attendance at meetings of the Académie Royale des Sciences. Normally an exclusively male space, the academy opened its doors to women, and other men as well, on an occasional basis and presented generalized versions of their work in progress. The presence of women, mostly nobles, helped legitimize the work done by the academy while at the same time, as historian Mary Terrall has argued, the gender divide widened between male academician and a female or feminized audience.[15] Women, however, could take advantage of other opportunities to gain access to and appropriate natural philosophy. Popularizers recognized the potential of this audience and began to actively seek out women as auditors for their lectures and demonstrations.

Women appeared prominently in the advertisements for these classes. Throughout the eighteenth century, popularizers repeatedly emphasized the dictum that science should appeal to "both sexes." The popularizer Allard, for example, offered several courses on experimental physics. In fact, in 1763, encouraged by the popularity of his current course, he added a second section. The first occurred in the mornings while the new addition took place in the afternoons. His goal was to join knowledge of physics with both the "useful arts and fashion." As such, he focused on the principles of mechanics, hydraulics, optics, astronomy, and clock making with the goal of putting this information "at the level of people of one and the other sex."[16] The Abbé Famin echoed this sentiment, that both sexes could and should attend public science courses; for his part, he offered free courses in experimental physics throughout the 1780s. He claimed that the people of Paris had become infatuated with physics and so he offered his classes, up to a maximum of fifty people per session, in order to meet the needs of the public. Famin claimed that his courses, which met three times each week for a total of forty classes, were models of clarity. As such, he "believed that auditors of both sexes" could easily follow and understand the content of experimental physics classes and that all of the men and women who had followed his course over the previous years had not done so without some benefit.[17]

Popularizers like Allard and Famin, however, offered courses that met several times each week over several months. At the other end of the popularization spectrum we find courses offered for one day only. These classes placed somewhat more focus on entertainment than education, although they still offered some access to a deeper understanding of experimental physics. More importantly, they utilized the same language with respect to their perceived and potential audience. François Pelletier, for example, operated in and around the Paris fairs and the Boulevard du Temple from the 1760s well into the 1790s. In 1785 he offered two versions of his course, one specifically for women. During the French Revolution he wrote that he hoped that his demonstrations would both instruct and amuse, and asserted that "ladies might equally enroll" and attend his lectures.[18]

A question over the content of the lectures arises from this willingness and even eagerness to have students of both sexes attend popular science courses. At a time when formal women's education in science hardly existed, it seems possible that popularizers watered down coeducational classes. This certainly seems to be the case with the course offered by Manoury who set up his demonstration laboratory next to the Café de Foi on the rue de Richelieu. He wrote that his lecture series, composed of twenty lessons covering subjects such as astronomy and geography, particularly would please ladies since he had stripped his courses of the abstract and used only concrete instruments, such as an astronomical sphere, to demonstrate science. In doing so, he explicitly invoked the precedent of Bernard de Fontenelle, whose *Conversations on the Plurality of Worlds* had so elegantly disseminated a Cartesian worldview to so many people.[19] Fontenelle's book had launched a trend in scientific popularization towards a female audience. Numerous books, many modeled directly after the *Conversations* and utilizing the dialogue format, appeared throughout the eighteenth century. These books maintained a steady popularity and often went through multiple editions. Rosnay's *La Physique des dames* appeared in 1773 while a decade earlier an author named de Rancy had offered an "essay on physics in the form of letters, for the use of young people of both sexes." This book focused on such topics as magnets, astronomy, and electricity.[20]

On the other hand, books written explicitly for a female audience did not always dilute the information found within them. Sigaud de la Fond's *Physique générale et physique particulière*, which appeared in five volumes from 1788 to 1789 as part of the *Bibliothèque Universelle des Dames*, did not hold back in his explanations of physics; instead, he simply avoided overly mathematical proofs in a manner similar to that in other books of the type intended for men and women. In place of mathematics, Sigaud de la Fond concentrated on experiments and demonstrations that readers who had instruments and equipment of their own could analyze and even reproduce.[21] Similarly, we can find evidence that not all public lecture courses overly simplified the information they were providing merely because they were trying to get women to be part of the audience. Jean-Antoine Nollet acknowledged that his audience was of mixed gender, but still aimed to disseminate his unique take on science along with other theories.[22] Allard wrote that people should not be "afraid" to invite ladies to a physics class; instead, his courses presented an excellent "educational plan" for women to follow and one that would "familiarize them with best quality of knowledge" through a method that the women would fine "agreeable and useful."[23] While many popularizers focused on the experimental rather than the theoretical, this does not imply that their courses were less advanced. Instead, disseminators based their decision on pedagogical and commercial reasons. When approaching a group of students with little, or in the case of women probably no, formal background in the sciences, a good way to begin the learning process was through visual demonstrations and explanations rather than abstract theoretical discussions. Just because they removed the mathematics and based the courses on ingeniously inventive and entertaining experiments did not mean that popularizers ignored the serious content in science, although it does imply that these courses frequently were set at an introductory level. On the other hand, these courses could allow one to watch the fun and ignore the explanation, a situation that surely arose on occasion. But we cannot determine how the audience viewed the courses since descriptions from the point of view of auditors in general, much less female auditors, are rare. Commercially, it makes sense that popularizers targeted as broad an audience as possible.

The mixed attitude toward the idea of teaching women about experimental physics, along with the related question of whether or not to dilute the information, exemplifies the general debate known as the *querelle des femmes*, concerning the educability of women and their place within the world. The quarrel over women's ability to reason resurfaced in France during the seventeenth century and produced some works in favor of that notion, most significantly François Poullain de la Barre's well-known *On the Equality of the Two Sexes*. Such works did not sway everybody, however, and the debate raged on throughout the Enlightenment.[24] Rousseau, for his part, claimed that people should not, and probably could not, educate women in the sciences. Rousseau believed that although women might enjoy the study of physics, no matter how hard they tried, they would remember "only some idea of its general laws and of the cosmic systems."[25] Thus, the range of approaches taken by popularizers of experimental physics mirrored the ongoing debate over the possibility and direction of female education.[26]

Illustrations of this complex set of attitudes towards women attending popular science courses appear within the plethora of comments popularizers make regarding their presence in the audience. In some cases, for example, popularizers designed special experiments or performed specific demonstrations just for their female auditors. As mentioned earlier, they repeated the experiment called the "electric kiss" to great acclaim and suggested that women, in particular, enjoyed it. This, perhaps, elucidates both John Adams's explanation of sexual attraction as well as the noblewoman's response. Experiments and explanations equating love to science also appear in Jean-François Pilâtre de Rozier's lecture on Newtonian physics, where he equates gravitational attraction to love. François Bienvenu, an instrument maker and physics demonstrator, offered short courses for several years before "adding a great number of very curious [experiments] on account of the ladies who desired to follow his course." On the other hand, the Abbé de la Pouyade felt that his entire course of physics was "the physics of ladies," in addition to the rest of the world, due to the intelligible manner in which he taught. Perrin developed specific physics "tricks" for his women auditors.[27]

Perrin's experimental physics classes bordered on the charlatanesque and hovered near the ever-shifting line between science and pseudosciences; however, his shows generally contained more science than not. If anything, he transformed actual scientific experiments into demonstrations that appeared magical and spectacular. In some ways, this defeats one of the purposes of the classes, that of making scientific knowledge explicable and potentially useful. Nonetheless, Perrin was mostly guilty of pushing the entertainment aspect of science a little too far. Other popularizers took explicit steps over the line between science and magic. Such an individual was Charles Rabiqueau, a former lawyer who transformed himself into professor of "occult physics" and whose courses ran the gamut from experiments on electricity and optics to magic. On the more scientific end, Rabiqueau frequently repeated the famous experiment done by many before him of having a group of people, all holding hands, receive an electric shock from a Leyden jar. Rabiqueau transformed this demonstration, however, and drew the subjects for the experiment from his audience.[28] On the occult end of his experimental repertoire, we find more specific references that focus on women. Rabiqueau claimed that women loved and desired his brand of occult physics, replete with amusing and recreational tricks and experiments. As such, he set up many experiments along these lines for his audience and even calculated the horoscopes of the ladies present, an act that pushed him from pseudo-magical physics into the borderlands of astrology.[29]

While some popularizers aimed specific experiments or demonstrations towards their female auditors, others went out of their way to ensure that the physical environment of their *cabinets de physique* were appropriate for women. François Pelletier emphasized this, for example, and asserted that "women would never suffer" at his courses from "the annoyance of the cold" because of the precautions he had taken for heating against the rigors of the winter weather.[30] While Pelletier concentrated his attention on climate, Paul Filidort assured his potential students, both female and male, that he would not force them to attend dangerous demonstrations and that they would not have to endure any noxious odors.[31] Comments like these lead us to consider the status of the women who attended popular lectures courses. As we saw above, Nollet's contemporaries described him as having the carriages of duchesses and other noblewomen lined up outside his door. Similarly, when advertisers mentioned women, they usually described their potential auditors as "ladies" [*dames*] rather than more simply and generally as "women" [*femmes*]. Conceivably, this fit in with the environmental concerns addressed by the popularizers themselves who likely targeted those women whose days generally passed without worry about either weather conditions or extreme odors.

The consumption of science

Public lecture courses emerged as a commodity in eighteenth-century France. These courses provided a great number of people with a broad access to science. With exceptions such as the free courses offered by the abbé Famin and a few others, however, we should note that these classes did come at a price. The consumption of goods rose in the eighteenth century throughout Europe, a phenomenon that included the purchase of food items such as coffee and tea as well as material possessions like stockings, pocket watches, and umbrellas. Science and medicine had a significant place in this trend. While much of this buying and selling fell within the realm of material goods, a market also existed for the ephemeral. In other words, we can identify a distinct market for the buying and selling of knowledge. While true for Europe in general, this phenomenon played an important part of the commercialization of enlightened ideas in France.[32]

Popular science traversed the spectrum of market behavior and provided interested individuals with a variety of items that they might purchase for their scientific education. Some of these commodities fall into the realm of the material. Women certainly bought books; as noted above, popularizers wrote numerous tomes especially for a female readership. Many of these books, such as Nollet's *Leçons de physique expérimentale*, were designed in such a way that the experiments could be reproduced by the reader at home. Thus, popularizers encouraged their readers to purchase scientific instruments. They might purchase the portable experimental physics kit sold by François Bienvenu, for example. In his advertisement, he guaranteed the kit would fend off ennui during long visits to the countryside, an ailment suffered by the elite, and especially women, but probably not by the lower classes. While Bienvenu did not specify the intended gender of his potential customers, the sales pitch would seem to include both men and women.[33] Although people could purchase an entire set of instruments, it more frequently happened that someone might purchase just one or two items with which to perform experiments. Bienvenu makes a point of mentioning his solar microscopes in the same advertisement in which he emphasizes the number of experiments he had created specifically for women. Clearly he hoped to sell a few microscopes as a result of his courses.[34]

Some nobles owned instruments. Madame de Pompadour, for example, had more than fifty scientific instruments, models, or machines including many recommended by Guyot in his *Nouvelles récréations physiques et mathematiques*.[35] Other women had instruments they no longer needed and wanted to sell. The Comtesse de Raigecourt in Nancy, for instance, advertised the sale of a solar microscope and a pocket telescope in a local periodical.[36] A female jeweler, who did not give her name, put a large electrical machine up for sale in Paris although the advertisement does not specify whether she had owned and used the machine, had purchased it just to resell, or acted as an agent for a third party.[37] Jacques Bianchi, a popularizer and instrument maker, gained considerable fame as well as the approbation of the Académie Royale des Sciences and the Faculté de Médecine for a breast pump he invented. While this does not quite fall under the category of an experimental instrument, the marketing of this medical device may have led to an increase in the number of women who entered his shop. He certainly targeted some women, namely midwives and expectant mothers, as the suggested consumers of his product. Thus, it is possible that he translated his development of a breast pump into an increased business overall with female clients.[38]

The purchase of scientific equipment, however, played only one part of the overall market in popular science. More important were the public lecture courses themselves. The cost of lectures varied from place to place, but not as much as might be expected. On a per-lecture basis, the courses offered around Paris cost between two and three livres. The difference largely lay in the length of the courses being offered. Brisson, for example, charged seventy-two livres for his thirty-six-lesson course, with each lecture lasting about two hours. At two livres per lecture, this admission fee was actually a little below average, but the overall cost was quite high and certainly precluded any but the fairly wealthy from attending. Thus, Brisson did not attempt to disseminate his knowledge too widely, preferring instead to lecture to the financially secure.[39] A reasonably well-off artisan might make ten to fifteen livres each week; thus it seems possible that some of his income might go for a single lecture or demonstration; but lengthy courses clearly fell outside the realm of the economically possible for most artisans. With this in mind, we might safely conclude that only the nobles and the upper middling sorts – well-to-do shopkeepers, merchants, professionals, and other members of the "bourgeoisie" – could afford the lengthier courses. Nollet's courses provided the exception to this rule; because of his position at the College Royal, his courses were, by default, free. However, some of the popularizers who offered lengthier courses also offered deals for their auditors. Pagny's class, consisting of eighteen lectures over six weeks, normally cost three livres per lecture. However, he offered a discount for his more loyal students; those who enrolled for the entire six weeks received a discounted price of thirty-six livres, a savings of thirty-three percent of the regular lecture price.[40]

Admission to the courses offered near the Palais Royal cost about three livres per lecture at most, just like many of the classes offered by those popularizers working on the Left Bank. The difference tended to be the length of the course, which with a few exceptions (such as Charles) were shorter. Marat's eight-lecture course on his theory of fire, for example, cost twenty-four livres. It also cost three livres per lecture to attend the courses given by Greppin and Billiaux, Bianchi, and Bienvenu. Not everybody adhered to this general rule; in fact, some demonstrations could cost quite a bit less. Bianchi and Bienvenu, for example, also offered various subscription packages to their *cabinets de physique*. Bianchi's subscription allowed individuals unlimited access to his demonstrations for twelve months. Since he usually offered several courses in a year, lasting typically one month each, this could be quite a bargain. In addition, subscribers saw and tried out all of the latest machines and instruments. Bienvenu, on the other hand, offered a deal whereby subscribers could come and go to any of his demonstrations as they wanted, but still had to pay for his more formal courses. Periodically, Bienvenu offered free courses on particularly popular topics. After the invention of ballooning, for example, he offered a six-lecture course on the nature of gases and electricity, and their relationship to aerostatics, at no cost, presumably to lure prospective customers to his shop.[41]

The shows and demonstrations found on the Boulevard du Temple could cost considerably less, if only because they tended to be one-night-only affairs. Popularizers who worked out of the theaters, for example, typically adopted the price scale already in use there; as such, a seat in the galleries could cost as little as six sous (0.3 livres) while a box seat might cost twenty-four sous (1.2 livres). Paysan de Nort-Hollande's demonstration at the Saint-Germain fair followed this pattern as did both Val's class at the Théâtre d'Emulation and Perrin's at the Théâtre des Associés. These prices did not go up until about the time of the French Revolution when the range went from twelve sous to three livres.[42]

For some popularizers, the price varied depending on the time of day. For those able to attend the 11:00 a.m. to 1:00 p.m. sessions of Pelletier's class, the cost typically ran to three livres. The evening shows, however, lasting from 5:00 p.m. to 7:00 p.m., cost just over one livre per person. This indicates that Pelletier assumed that his audience during the day would be able to pay a higher admission fee that his nighttime audience, as the latter presumably consisted of individuals who had to work during the day and had a lower income. He also advertised group discounts; a company of five people, for example, could enter his shows for only six livres.[43]

The price of lectures, then, depended on location. Popularizers also made some alterations in the price structure exclusively for women, perhaps as a way to expand their potential consumer base. The lectures that Manoury offered exclusively for women had the added inducement of allowing the fathers, mothers, tutors, and governesses of the auditors to attend the course for free. Clearly, Manoury had designed his lecture series for well-off young women whose parents sought to provide them with an education either through their own hands or by hiring someone to teach their daughters for them.[44] The popularizer Brogniart illustrates another alteration in the price structure that favored women; he offered a class in physical and medical electricity in 1791 that cost twenty-four livres. Women, however, could attend his course for half price. In some ways this can be seen as an inducement to get more female auditors. Alternatively, it could be viewed in negative light. Just as women workers often received lower wages because employers assumed that they could not work as hard, it is possible that Brogniart charged less because he felt that women would learn less. One bit of evidence that points more towards a positive reading of this price structure comes from an examination of some of the membership fees for the musées that appeared in the 1780s and 1790s. It cost women half as much as men to join the Musée de Monsieur, for example.[45]

Certainly, women had a strong presence among the many customers of popular science, both in the purchase of material objects like books and scientific instruments and in the more intangible purchase of cultural capital acquired through one's attendance at public lecture courses. However, we should note that occasionally women acted as the purveyors of scientific equipment and knowledge. When Jacques Bianchi died, for example, his widow took over the shop and continued to sell scientific instruments along with his famed breast pump. Women occasionally published books that included popularized science. The female author of the *Traité d'éducation des femmes, & cours complet d'éducation* included a section on experimental physics, based largely on the works of Nollet.[46] In rare instances women might even teach a course in experimental physics. In 1790 Madame du Piery opened a course in astronomy that met three times each week for a total of twenty or twenty-four sessions. She planned to explain the theory and the practice of astronomy and to lead her auditors in an examination of the constellations and the planets.[47]

The geography of popular science

Examining the locations for the dissemination of enlightened science within the emerging public sphere allows us to trace the changing topography of scientific appropriation over the course of the eighteenth century. This geographic expansion can be tied to the growth of consumer culture and the development of scientific knowledge as part of the cultural capital many people of the period wanted to acquire. Outlining the interplay between different aspects of the spectrum of scientific popularization elucidates the overall shape of popular scientific culture in Paris. Although historians frequently present science as significant for the elite, we must recognize the important presence of science within urban culture more generally and acknowledge the changes wrought upon science through its commercialization. In particular, the spread of science is best viewed via the lens of the developing public sphere. Just as books, periodicals, and the various institutions made knowledge and information available to more and more people, so too could the ideas of the Enlightenment be appropriated orally. Public lecture courses provided an entry point into this cultural milieu; thus, the appearance of such courses in different parts of Paris over the course of the century marked the transmission and reception of enlightened ideas by different audiences at different times. The factors in this analysis – including audience, science, and geography – reveal the interconnections between different parts of urban culture and the ways in which the urban population participated in the creation of that culture.[48]

The three main locations for lecture courses in Paris were the Left Bank, the region on and around the Boulevard du Temple, and the area near the Palais Royal. These locations developed over several decades, in the order listed, and grew to meet an increase in consumer demand. The beginning point, on the Left Bank, clearly operated within the parameters of the educational system found in and around the Sorbonne and the Paris colleges. Popularizers worked alongside and occasionally within this educational system and provided people, mainly students but others as well, with a much-needed experimental view of natural philosophy. This education was mirrored, if somewhat hazily, in the subsequent growth of interest in popularized experimental physics in the entertainment sector of the Boulevard du Temple. Disseminators, working among the theaters, cafes and other shops, provided a cheaper, shorter version of experimental physics that targeted an audience more inclined towards amusement than education. A more upscale version of these entertaining, but also instructive, lectures then moved over to the higher-rent district near the Palais Royal and along the rue Saint-Honoré. This growth took several decades and illustrates the geographic dissemination of public lecture courses through the streets of Paris.

The Left Bank held multiple opportunities for individuals seeking to learn about experimental physics. The audience for the popular science found there came from the educated and the elite. These were people with leisure time and money to spend, occasionally, on lengthy courses of a relatively sophisticated nature. Both men and women could be found here although the main clientele were male students.

Chronologically, the area around the Paris colleges was the first of several sites where popularization activities concentrated. Experimental physics had been part of secondary education in Paris since the beginning of the eighteenth century. The efforts of Pierre Polinière, for example, ensured that the science taught at these institutions reached students through a variety of methods. It was not, however, until Louis XV created the post of professor of experimental physics for Jean-Antoine Nollet at the Collège de Navarre that individuals who specialized in demonstrations began to enter the formal educational system. Beginning with Nollet, several popularizers held jobs in the various colleges and technical schools. Nollet's classes were so popular it was reported that as many as five hundred people would attend each class. In some colleges, especially the Collège de Navarre and the Collège Royal, classes were open and free to the general public. Professors who held educational positions, like Mathurin-Jacques Brisson, Jean-Antoine Nollet, and Louis Lefèvre-Gineau, taught courses designed for the general public.[49]

Most of the popularization efforts around the Left Bank, however, were not due to the efforts of the official teachers but to people, like Polinière, who worked as demonstrators. These individuals performed experiments for the college students and offered courses on their own time, perhaps with the hope that a more lucrative professorship might become available. Most waited in vain, but a few were rewarded for their patience as well as for their talents. Allard, for example, got a job teaching mathematics at the Ecole Royale Militaire and Mathurin-Jacques Brisson, who started out as a demonstrator for Nollet, eventually took over his post at the Collège de Navarre. Joseph Aignan Sigaud de la Fond, who also spent many years working for Nollet, failed to acquire any of Nollet's posts. Nonetheless, his series of courses were extremely popular and, when he retired, he passed on the business to his nephew, Rouland, who continued to practice into the period of the French Revolution.

By the 1740s there were several disseminators on the Left Bank offering lectures and demonstrating the principles of natural philosophy for the benefit of students, among others. A certain Delor, for example, began advertising his classes on electricity in 1746. A protégé of Georges Louis Leclerc, comte de Buffon, Delor earned his reputation through the facility with which he reproduced certain experiments developed by Benjamin Franklin concerning electricity and lightning rods. Delor found steady work as a popularizer through the 1760s and gained a loyal following for the numerous courses in experimental philosophy offered in his *cabinet de physique*. Ironically, his contemporaries equated his popularization efforts with those of his rival, Nollet, and his later courses even utilized Nollet's textbook, the very popular *Leçons de physique expérimentale*.[50] In addition to his private lectures, Delor gained two positions as an official demonstrator with various colleges near the University of Paris and at the Ecole Royale Militaire.[51]

While science earned a place within the educational confines of the Left Bank region, other geographic locations also experienced the development of a conglomeration of scientific enterprises. The main region for this growth was the Boulevard du Temple, which had long functioned as a center for popular entertainment. It "offered spectacles from every genre: besides the comedy theaters, one finds conjurors in their rooms [and] a barker straining himself to announce experiments in physics, card tricks, etc."[52] This famous Paris street attracted a socially eclectic crowd that congregated there for a number of reasons. The streets and boulevards of Paris had always been home to individuals plying their trades. Some of these tradesmen grew famous for their work, such as the notorious dentist "le grand Thomas" who pulled teeth and entertained crowds in the shadow of Henry IV's statue on the Pont-Neuf.[53]

The Boulevard du Temple, along with the Paris fairs, attracted enormous crowds thanks to a dense concentration of entertainments. In addition to well-known cafés and shops, this street was home to early wax museums and several theaters that drew a large and diverse audience. Although street theaters were scattered all over Paris, the Boulevard du Temple contained the most famous ones such as the Ambigu-Comique, the Théâtre des Associés, and the Variétés-Amusantes. These theaters anchored the area as a destination for individuals seeking amusement. In and around these theaters were scientific popularizers who performed demonstrations and sometimes offered more extended courses in experimental physics. Men and women from all social classes, the elite included, could easily penetrate the more popular milieu of the Boulevard du Temple.[54] Everybody could participate in the entertainments found on the Boulevard du Temple and the prices there remained the lowest in Paris.

A number of people established themselves more or less permanently on the Boulevard du Temple. Zaller, for example, set up his *cabinet d'optique* in various places around Paris before he settled down near the Ambigu-Comique in 1770. He demonstrated the possibilities of optics by creating various illusions representing such things as public festivals and fireworks. In order to accommodate the greatest number of people, Zaller kept his cabinet open from ten in the morning until ten at night. In 1779 Noël constructed a *cabinet de physique* on the boulevard near the famous Café Turc. From his shop, he gave demonstrations of various instruments and machines that were for sale, including a portable collection for people who wanted to amuse themselves around Paris or in the countryside. At any time when his cabinet was open, usually from eleven in the morning until eleven at night, he might present demonstrations for individuals or groups seeking special training. These were on a broad range of subjects, including electricity and optics, and were always, so he claimed, recreational and entertaining. Noël continued working into the revolutionary period. For a while he teamed up with Val to put on a show at the Cirque National in the Palais Royal, although this seems to have been their only foray into that part of town.[55]

Some popularizers worked exclusively in the theaters on the Boulevard du Temple. The popularity of these theaters probably aided in attracting an audience for the scientific shows produced there. Unfortunately, little is known about the inner workings of these small theaters. It is unclear, for example, how many people they could seat. Also unknown exactly is the relationship between the disseminators and the theater owners who allowed popular science demonstrations between the staged productions. Perrin, for instance, frequently worked out of the boulevard theaters, usually for several days in a row. In 1787 he performed his "amusing physics" show out of the Théâtre des Délassemens Comiques, while in 1789 he used the stage at the Théâtre des Associés.[56]

Louis-Sébastien Mercier claimed that the people of France called the Palais Royal "the capital of Paris"; true or not, it certainly offered a unique cultural atmosphere. Located near the Louvre and the elegant homes and shopping found on streets like the rue de Saint-Honoré, the district around the Palais Royal clearly catered to a social and economic elite. The garden of the Palais Royal was a popular promenade for high society by day and for prostitutes and their clients by night. In spite of its location in an upscale neighborhood, the inner courtyard of the Palais Royal, sometimes referred to as the Palais Marchand, featured a wide variety of shops, cafés, and boutiques catering to multiple levels of society. This cultural milieu also included the shops and cabinets of scientific popularizers. Their presence was well established and no contemporary description of the Palais Royal failed to mention the available physics courses or some of the scientific wares available for purchase there.[57] In the guidebook for visitors to Paris written by Luc-Vincent Thiéry, for example, the author notes several boutiques that specialize in the sale of scientific instruments as well as scientific clubs and courses.[58]

There is a clear chronological progression concerning the availability of popular science courses around the Palais Royal. Unlike the Left Bank, where these courses began to appear during the 1730s, and the Boulevard du Temple, where they began in the 1750s, only a few individuals offered classes at the Palais Royal before the 1770s. In the 1750s, for example, Triboudet de Mainbray, a doctor of civil and canon law from Great Britain, taught an experimental physics course in his *cabinet de physique* near the Palais Royal on the Place des Victoires. Triboudet de Mainbray, whose family had fled France after the revocation of the edict of Nantes, operated one of the best cabinets in England, or so he claimed, and enjoyed membership in numerous scientific academies such as those of Bordeaux, Lyon, Toulouse, and Montpellier. Having abandoned England for France, he established himself near the Palais Royal and began offering a series of thirty-four lectures covering mechanics, astronomy, optics, and magnets.[59]

Another course, this one available in the late 1760s, was made possible by the abbé de la Pouyade, an ardent proselytizer for the physical system created by the amateur physicist Jacques-Charles-François de la Perrière de Roiffé. La Pouyade summarized this system in a three-month series of lectures covering both experimental and general physics. La Pouyade felt that La Perrière de Roiffé's system had "the advantage of being simple and at the level of everybody." Despite this claim, his notion of physics and the natural world never really gained much acceptance.[60] Aside from a few exceptions, scientific disseminators did not establish operations around the Palais Royal until the 1770s and 1780s, well after popularization had become a fixture on the Left Bank and the Boulevard du Temple. The fifteen years before the Revolution, however, saw a dramatic increase in the options available to potential auditors. Girardin, for instance, a self-styled professor of physics, offered courses out of his *cabinet de physique* located in the garden of the Palais Royal itself. His twelve-lesson class covered electricity, the nature of electric fluids, and the theory of aeriform gases. Upon request Girardin would repeat any particular experiment after the class was over.[61]

Manoury established his cabinet beside the Café de Foi, a venerable establishment located in a passageway between the rue de Richelieu and the Palais Royal. There he offered a wide variety of courses on astronomy, physics, and mechanics. These classes presented a generalized range of knowledge filled with numerous practical demonstrations and experiments using, among other devices, an electrical machine, a reflecting lamp, a singing automaton, and a hydraulic fountain. The demonstrations he performed with some of these machines gained fame independent of his experimental physics course. For example, he displayed the singing automaton on its own in a *cabinet de physique* next to Audinot's theater on the Boulevard du Temple. Despite this association with boulevard entertainment Manoury also claimed a direct connection between his course and the work of elite savants. "The method adopted in the course," he stated, "presents to the eyes and to the mind a kind of recreation, which gives everybody at the time some enlightenment and some very satisfying knowledge of the newest and most important discoveries, according to the observations of the Académie Royale des Sciences."[62] Manoury offered his audience a quick and painless foray into enlightened science.

Not all of the public courses offered around the Palais Royal were so accepting of the work of the academicians. Jean-Paul Marat, who worked out of his *cabinet de physique* on the rue Saint-Honoré, spent several years trying to convince members of the Académie Royale des Sciences, and the public at large, that his own theories regarding light and optics should supersede those of Isaac Newton. Among other claims, he stated that Newton's views on refraction and the primary colors were incorrect. Marat had given up his career in medicine in order to focus attention on what he considered to be incredible discoveries in the field of physics. The fact that Marat could not convincingly prove his theories to the commission sent by the Académie to examine his claims did nothing to bolster his position within the scientific public sphere. Despite Marat's audacity in publicly attacking Newtonianism, an act equal to criticizing the Enlightenment itself, his scientific efforts are actually on a par with much of the scientific activity in Paris in the eighteenth century. Most of the modern historiography, however, focuses on Marat's attacks on the Académie Royale des Sciences after the latter rejected his theories, rather than on an examination of his actual work.[63]

Marat did receive some criticism from his contemporaries, both elite savants and fellow popularizers. Jacques-Philippe Ledru and Pierre Bertholon, for example, criticized Marat's work on electricity and electric medicine.[64] Without a doubt, the rival most hated by Marat himself was Jacques-Alexandre-César Charles, a experimental physicist of considerable renown. Charles, during one of his lectures, openly criticized Marat's theories. When word of this reached Marat, he went to Charles's house and demanded an apology. Charles refused to recant and the two ended up fighting in the mud. Ultimately, they challenged one another to a duel, an event stopped only by the Paris police.[65] This episode emphasizes the fragile nature of one's scientific reputation and the lengths an individual might go to in order to maintain an air of respectability and the good graces of public opinion.

Charles himself first started giving lectures in experimental physics in 1780 after losing his job in the office of the controller-general. His lectures, held in his *cabinet de physique* on the Place des Victoires, enjoyed enormous popularity throughout the 1780s and 1790s. Charles convinced his auditors that "without doubt, nothing is more interesting than the spectacle [of experimental physics]." During the French Revolution, in addition to his classes, Charles worked for the new government doing inventories of the scientific instrument collections of the émigrés. His own personal collection of instruments was the largest and "most beautiful" *cabinet de physique* in Europe. Charles's classes attracted amateurs and savants alike: both Franklin and Volta attended his lectures while visiting Paris. He gained even more fame in 1783 when he became the first man to ascend in a hydrogen balloon. Unlike some of his colleagues, and thanks to the work he had performed for the revolutionary government, Charles was able to translate his success as a popularizer into economic security and academic prestige. When the Institut de France rose from the ashes of the Académie Royale des Sciences, Charles found himself elected one of its first members. Although he never published any of his work, Charles gained fame throughout Europe for his abilities as a demonstrator. His course in physics frequently appeared as a chief attraction in contemporary guidebooks. He offered several different classes, including a complete course in experimental physics in thirty-six lectures and another course in electricity. His chief course, however, contained sixty lectures and covered topics in mechanics, optics, acoustics, chemistry (usually related to the nature of air and other gases useful for aerostatics), and electricity.[66]

Charles's classes were but one among many options available during the 1780s. In 1789 Bouthoux de Lorget, a self-styled mechanic and physicist, opened his *cabinet de physique* in the Palais Royal itself. His two-and-a-half hour performances took place in the Théâtre des Beaujolais and included a large number of experiments using his collection of physics instruments and machines, as well as demonstrations of his automatons. Between performances, Bouthoux de Lorget executed many "surprising tricks and physics experiments," apparently in an attempt to attract an audience. Earlier in the decade two enterprising physicists and instrument makers, Greppin and Billiaux, opened a short-lived *cabinet de physique* where they performed experiments designed to illustrate all of the diverse and interesting parts of physics. Although the work as demonstrators of science was short-lived, they did continue to make scientific instruments and demonstrate their use throughout the 1780s.[67]

In 1786, the abbé Pierre-Noel Famin began offering a free course in experimental physics at the Hôtel of the duc d'Orléans near the Palais Royal. "In our time," he wrote, "physics has become a more interesting science than ever before. Each day it makes new progress; each day it offers us new phenomena, more instructive than what has come before." It is no wonder, Famin thought, that "everybody studies [physics]." He believed that the desire to learn gave honor to the century of the Enlightenment and to France in particular. "It is in France, it is in that capital where foreigners abound, that the *lycées* were established, [and] that the professors multiply."[68] He continued giving his courses for the next eight years until, apparently, he had a change of mind and switched to a career as an author and dramatist.

If Famin's general course did not interest prospective students, they could always attend some of the more specialized classes offered by Jacques Bianchi in his *cabinet de physique* on the rue Saint Honoré. Bianchi, an Italian instrument maker with more than thirty years' experience, performed demonstrations in his workshop and offered a series of short courses, usually between four and eight lectures, on topics such as electricity and the inner workings of microscopes. Like Charles, Bianchi had some connections with the more elite savants of his time who, according to his own account, visited his shop to consult him regarding different electrical machines.[69]

François Bienvenu, another popularizer who first appeared during the 1780s, played a double role by making and selling scientific instruments in addition to offering classes in experimental physics. His instruments, modeled after those of Nollet, Joseph-Aignan Sigaud de la Fond, and Guyot, were sold all over France and included everything necessary for a complete *cabinet de physique*. His short courses, usually four to six lessons on topics like electricity and light, were clearly designed to show off his instruments. At other times he even gave free demonstrations; one such course examined the nature of gases, probably in order to capitalize on the popularity of ballooning and thereby to draw more people into his shop. Bienvenu also tried to exploit the balloon craze by publishing a short pamphlet outlining several machines he and another savant had contributed to the field of aerostatics.[70]

The Palais Royal contained a fairly extensive variety of courses and shops geared toward providing knowledge of experimental physics to their customers. This area offered much to the interested amateur and the savant alike; new machines at Bianchi's shop drew individuals from multiple levels of society, as did Charles's lectures. The range of scientific activity in that area offered the greatest mixture of elite savants and lower-level entertainers and a wide range of forums and commercial venues from *cabinets de physique* to theaters and cafés. Overall, the wide variety of locations for scientific popularization, from the Left Bank to the Boulevard du Temple and around the Palais Royal, combined to make natural philosophy truly accessible to the bulk of the Parisian population.

Interconnecting geographies

Just because popularizers started out in one part of town did not mean they could never move to a different region. In fact, some popularizers exerted a considerable amount of effort in relocating to different parts of Paris. Moving to a different part of town is a simple matter of geography; however, it does not make one accepted there. For that, more work is required. Some examples of the difficulties inherent in moving from one part of Paris to another will help illustrate this point. Nicolas-Philippe Ledru began his career as a popularizer on the Boulevard du Temple in the 1750s, where he catered to the "epidemic curiosity" of Parisians for physics experiments and games. Not all of his work was exclusively intended to entertain his audience. He also conducted experiments in physics with an eye towards improving the state of knowledge in that field. In his own words he hoped to fulfill a dual function to both "excite the curiosity of the people and extend the progress of physics."[71] Ledru wanted to apply his knowledge of experimental physics, and most particularly of electricity, to the general benefit of humanity, while at the same time striving to improve his position in the scientific public sphere.

Ledru wasted little time gaining noble patronage. By 1769 he had performed some of his experiments for Louis XV. From 1773 to 1776, he won the patronage of the duc de Chartres and would often perform electrical experiments and demonstrations for his benefit. Ledru earned a certain amount of respect and admiration from his colleagues at that time and his experiments were often reported in the *Journal de Physique*. In 1783, Ledru applied for and received a license from Louis XVI to practice the trade of medical electrician at his new Hospice Médico-Electrique located on the Left Bank of Paris, a significant geographical and conceptual leap from his previous locations in the Paris fairs and on the Boulevard du Temple. Ledru now worked full-time applying his knowledge of electricity to treat patients suffering from ailments such as epilepsy and paralysis, previously thought incurable. That same year, the Paris Faculté de Médecine investigated Ledru's activities and, initially, agreed that medical electricity as practiced by Ledru was an effective and useful treatment. By 1784 he had reached the pinnacle of his career and could use the title of "King's physicist" as well as claim the approbation of the Faculté de Médecine for his electrical treatments. It seemed that both the court and elite savants had accepted Ledru into their midst and that he could do no wrong. His smallest success was dutifully reported in the newspapers; in 1784, for example, the *Journal encyclopédique* claimed that he had cured a white Angora cat of epilepsy.[72]

Unfortunately, Ledru's rise was marred by a vicious attack on his reputation. A rival in the Faculté de Médecine, Pierre Jean Claude Mauduyt de la Varenne, who practiced a similar kind of electric medicine, tried to force Ledru out of business as a charlatan. Ledru survived the accusations and continued to practice during the revolutionary era, despite the tarnishing of his reputation.[73] The point here, however, is that Ledru navigated his way from the Boulevard du Temple towards the elite end of the scientific public sphere over the course of his career. Indeed, Ledru shows that while the journey from the Boulevard du Temple to the sphere of the elite savants may have been a long and arduous one, it was possible to succeed.

More successful in changing his geography was François Pelletier. He began his career working out of his *cabinet de physique* in the Paris fairs, until a fire at the Saint-Germain fair in 1762 destroyed most of his instrument collection. After several temporary locations, including one near the Pont Marie, the rue Charlot, and rue de Bretagne, Pelletier opened a new shop on the Boulevard du Temple in 1763. This shop was conveniently located next to the Café de Caussin, among the neighborhood's most famous. Pelletier, who described himself as an engineer and machinist as well as a physicist, gained considerable fame there for his demonstrations and innovative machines. The authors of the *Correspondance littéraire* claimed that Guyot, who had written a very popular book of experiments titled the *Nouvelles récréations physiques et mathématiques*, copied many of its demonstrations from Pelletier. In addition, the *Mémoires secrets* commented favorably on Pelletier's work with automatons. In all, Pelletier spent more than twenty years on the Boulevard du Temple, giving daily demonstrations of his machines, inventions, and other curiosities. As a result, he became a fixture in the popular science world. He supplemented his living by occasionally traveling through the provinces and abroad, performing demonstrations in his portable *cabinet de physique*.[74]

During the late 1770s Pelletier started wooing potential patrons in an effort to move up the social ladder. First, he obtained a royal privilege as a "king's engineer and machinist." This new title not only meant that he had purchased from the state a license to practice his craft, it also added a measure of prestige to his work. His next success came in the form of a royal patron, in this case the heir to the Spanish throne Dom Gabriel. In 1782, he published an *Hommage aux Amateurs des Arts*, wisely dedicated to Louis XVI, in which he outlined all of his achievements over the previous thirty years. Through this he clearly hoped to gain French noble patronage.[75]

At about the same time, Pelletier left his longtime home on the Boulevard du Temple and moved to the high-rent district on the rue de Richelieu next to the Palais Royal. By moving to a more fashionable neighborhood, Pelletier attempted to reach a more affluent and influential clientele. Well-off people had attended his shows in the Boulevard du Temple; but their presence could be more consistently counted upon in the Palais Royal. He also began to participate in other activities; he demonstrated a mirror polish for a scientific club in 1782. In his new quarters Pelletier offered more extended courses than he had given at his old location. Most of the spectacles on the Boulevard du Temple and in the fairs had been brief, one-day-only demonstrations. When he moved, Pelletier altered his teaching somewhat. Although entertainment was still the key, he now gave courses that typically lasted one month. In 1785, for example, he offered such a course that focused on the nature of magnets. Similar courses covered mechanics and hydraulics.[76]

The last step in Pelletier's geographical wanderings came in 1786, when he wrote a letter to the comte d'Angivilliers, the head of the king's household, asking for a royal pension. Unfortunately, the response is lost, but since Pelletier never mentioned earning a royal pension it is safe to assume that his request was denied. During the French Revolution, Pelletier, now in his sixties, continued to woo the government. Finally, in 1793, the members of the National Convention accepted his offer to donate his machines and instrument collection to the state in return for a stipend. His collection was to be housed in the national archives and he would live there as a caretaker. He continued to offer short classes on various topics and he created a Société des Arts. This society functioned like a small club focused entirely on the teaching of natural philosophy to its members who, Pelletier assumed, would be amateurs, artisans, and women. An annual subscription fee of forty-seven livres allowed members access to any and all of Pelletier's courses as well as to his *cabinet de physique*.[77] The geography of popular science changed and shifted alongside the growing public interest in appropriating natural philosophy.

Notes

1 John Adams, *Diary and Autobiography of John Adams*, ed. L.H. Butterfield, 4 vols. (Cambridge, MA: Belknap, 1961), IV:36–7.

2 *Affiches de province*, 11 June 1777, 95; *Avant-coureur*, 2 February 1767, 66, *Affiches de Paris*, 23 February 1767, 160 (on Sigaud de la Fond and Rabiqueau). On the problems of defining the audience see Steven Shapin, "The Audience for Science in Eighteenth-Century Edinburgh," *History of Science* 12 (1974): 95–121.

3 Nollet, *Leçons de physique expérimentale*, 6 vols. (Amsterdam & Leipzig: Arkstee & Merkus, 1754–1765), I:ix (at the level of everybody); ibid., I:xxxi–xxxii (young people).

4 *Affiches de Paris*, 3 February 1752, 78, ibid., 7 January 1754, 14; and ibid., 1 May 1755, 270.

5 Antoine Lavoisier, "Réflexions sur l'instruction publique," in *Oeuvres de Lavoisier*, 6 vols. (Paris: Imprimerie Impériale, 1862–1893), VI:521.

6 Nollet, *L'Art des expériences*, 3 vols. (Paris; P.E.G. Durand, 1770), I:xii. For Sigaud de la Fond see *Avant-coureur*, 9 November 1772, 713–14; ibid. 25 February 1771, 121; *Avis divers*, 22 February 1777, 235–6; *L'Année littéraire*, 1772, I:192; *Affiches de province*, 14 November 1781, 184; and *Journal de Paris*, 23 February 1782, 215. On Noël see *Affiches de Paris*, 27 November 1779, 2645; and on Miollan see *Affiches de province*, 30 March 1784, 188.

7 *Affiches de province*, 29 May 1784, 313 (Bienvenu). For foreign visitors see, for example, Bengt Ferrner, *Resa I Europa*, ed. Sten G. Lindberg (Uppsala, Almquist & Wiksells, 1956), 352; and Joseph Teleki, *La Cour de Louis XV: Journal de voyage du comte Joseph Teleki*, ed. Gabriel Tolnai (Paris: Presses Universitaires de France, 1943), 64 and 71. On guidebooks see Daniel Roche, ed., *Almanach parisien en faveur des étrangers et des personnes curieuses* (Saint-Etienne: Publications de l'Université de Saint-Etienne, 2001), 7–32; and Gilles Chabaud, "Les Guides de Paris: Une literature de l'accueil?" in *La Ville promise: Mobilité et accueil à Paris (fin XVIIe–début XIXe siècle)*, ed. Daniel Roche (Paris: Fayard, 2000): 77–108.

8 On Nollet's performance before the king of Denmark see the comments by the abbé de la Pouyade and J.-C.-F. de la Perrière, *Journal encyclopédique*, February 1769, 131–4; and ibid., July 1769, 271–80. On Joseph II and Ledru see Derek Beales, *Joseph II*, 3 vols. (Cambridge: Cambridge University Press, 1987), I:378. On Brienne see André Morellet, *Mémoires de l'abbé Morellet*, ed . Jean-Pierre Guicciardi (Paris: Mercure de France, 1988), 220. On the duc de Croy see Emmanuel duc de Croy, *Journal inédit du duc de Croy, 1718–1784*, 4 vols. eds Vicomte de Grouchy and Paul Cottin (Paris: Flammarion, 1906), II:319–20. On lectures for the Dauphin see Pierrette Girault de Coursac, *L'Education d'un roi: Louis XVI* (Paris: Gallimard, 1972), 209.

9 *Journal de Paris*, 26 September 1786, 1109–10.

10 Jean-Baptiste Le Roy to the comte de Tressan, 26 August 1749, in Louis-Elisabeth de la Vergne, marquis de Tressan, *Souvenirs du Comte de Tressan* (Versailles: Henry Lebon, 1897), 65–6.

11 Châtelet to Francesco Algarotti, 20 [April 1736], in Châtelet, *Les Lettres de la marquise du Châtelet*, ed. Theodore Besterman, 2 vols. (Geneva: Institut et Musée Voltaire, 1958), I:112. Algarotti, quoted in Barbara Stafford, *Body Criticism: Imaging the Unseen in Enlightenment Art and Medicine* (Cambridge, MA: MIT Press, 1994), 361.

12 On noblewomen in general see François-Marie Mayeur de Saint Paul, *Le Chroniqueur désoeuvré, ou l'Espion du boulevard du Temple* (London, 1782). On Sigaud de la Fond and his noble clients see Jean Torlais, *Un physicien au siècle des lumières* (Paris: Sipuco, 1954), 232–3; and Stephanie Félicité Ducrest de Saint-Aubin, comtesse de Genlis, *Mémoires inédits de Madame la Comtesse de Genlis*, 10 vols. (Paris: Ladvocat, 1825), II:334. On the comtesse de Sabran see Comtesse de Sabran to Chevalier de Boufflers, 25 April [1778], in *Correspondance inédit de la Comtesse de Sabran et du Chevalier de Boufflers, 1778–1788*, eds. E. de Magnieu and Henri Prat, 2nd edn (Paris: Plon, 1875), 5. On La Tour du Pin see Henriette-Lucie-Dillon La Tour du Pin, *Journal d'une femme de cinquante ans (1778–1817)*, 4 vols. (Paris: Chapelot, 1913), I:54.

13 Sigaud de la Fond, *Physique générale et physique particulière*, 5 vols., part of the *Bibliothèque Universelle des Dames* (Paris: Rue et Hôtel Serpente, 1788–1789), I:v.

14 On Rohault see Paul Mouy, *Le Développement de la physique cartésienne, 1646–1712* (Paris: Vrin, 1934), esp. 108–13; and Trevor McClaughlin, "Le Concept de science chez Jacques Rohault," *Revue d'histoire des sciences* 30 (1977): 225–40. On women and popular science in eighteenth-century France see Jeanne Peiffer, "L'Engouement des femmes pour les sciences au XVIIIe siècle," in *Femmes et pouvoirs sous l'ancien régime*, eds. Danielle Haase-Dubosc and Eliane Viennot (Paris: Editions Rivages, 1991): 196–222; and Peiffer, "La Litterature scientifique pour les femmes au siècle des lumières," in *Sexe et genre: De la hiérarchie entre les sexes*, eds. Marie-Claude Hurtig, Michèle Kail, and Hélène Rouch (Paris: Editions du Centre National de la Recherche Scientifique, 1991): 137–46. Bernard de Fontenelle, uncited quotation in Paul Dorveaux, "Apothicaires membres de l'Académie Royale des Sciences, VI: Nicolas Lémery," *Revue d'histoire de la pharmacie* 7 (1931): 208–19, quote on 211.

15 Mary Terrall, "Gendered Spaces, Gendered Audiences: Inside and Outside the Paris Academy of Sciences," *Configurations* 3 (1995): 207–32.

16 *Affiches de Paris*, 31 March 1763, 234.

17 *Affiches de province*, 30 November 1786, 580; and *Journal de Paris*, 11 December 1787, 1487.

18 *Journal de Paris*, 5 May 1785, 508 (course just for women); *Affiches de Paris*, 4 June 1790, 1574 (ladies enrolling).

19 *Journal de Paris*, 18 June 1777, 3.

20 Fontenelle, *Conversations on the Plurality of Worlds*, trans. H.A. Hargreaves (Berkeley: University of California Press, 1990). De Rosnay, *La Physique des dames, ou les quatre élémens; ouvrage utile pour disposer à l'intelligence des Merveilles de la Nature* (Paris: Stoupe, 1773); de Rancy, *Essai de physique en forme de lettres, à l'usage des jeunes personnes de l'un & l'autre sexes* (Paris: Hérissant, 1768). De Rancy's book is an augmented edition of a work anonymously published five years earlier titled *Lettres physiques, contentant les notions les plus nécessaires à ceux qui veulent suivre les Leçons expérimentale de cette science* (Paris: L.-G. de Hansy, 1763). On Fontenelle see Aileen Douglas, "Popular Science and the Representation of Women: Fontenelle and After," *Eighteenth-Century Life* 18 (1994): 1–14; and Michel Delon, "La Marquise et la philosophe," *Revue des sciences humaines* 182 (1981): 65–78.

21 Sigaud de la Fond, *Physique générale et physique particulière*.

22 Nollet, *Leçons de physique expérimentale*, I:xviii.

23 *Affiches de province*, 30 March 1763, 51.

24 On the debate over women in the Enlightenment see Lieselotte Steinbrügge, *The Moral Sex: Woman's Nature in the French Enlightenment*, trans. Pamela E. Selwyn (Oxford: Oxford University Press, 1995); and Sylvana Tomaselli, "The Enlightenment Debate on Women," *History Workshop Journal* 20 (1985): 101–24.

25 Rousseau, *Emile, or On Education*, trans. and ed. Allan Bloom (New York: Basic Books, 1979), 425–6.

26 On women's education see Samia I. Spencer, "Women and Education," in *French Women and the Age of Enlightenment*, ed. S. Spencer (Bloomington: Indiana University Press, 1984), 83–96; and Martine Sonnet, *L'Éducation des filles au temps des lumières* (Paris: Les Editions du Cerf, 1987).

27 The electrified boy could attract such items as bits of paper or metal shavings to his hand. See Nollet, *Essai sur l'électricité des corps*, 5th edn (Paris: Guerin, 1746), 76–82. Nollet did not create this experiment. It was first developed by Georg Bose, a German physicist who developed many novel demonstrations. See J.L. Heilbron, *Electricity in the Seventeenth and Eighteenth Centuries: A Study of Early Modern Physics* (Berkeley: University of California Press, 1979), 265–7; and Geoffrey Sutton, *Science for a Polite Society: Gender, Culture, and the Demonstration of Enlightenment* (Boulder, CO: Westview Press, 1995), 304–5. *Journal de Paris*, 7 July 1787, 825 (Bienvenu); *Avant-Coureur*, 29 December 1766, 823–5 (de la Pouyade); *Affiches de Paris*, 28 April 1793, 1815 (Perrin).

28 On this topic see, for example, Simon Schaffer, "The Consuming Flame: Electrical Showmen and Tory Mystics in the World of Goods," in *Consumption and the World of Goods*, eds. John Brewer and Roy Porter (London: Routledge, 1993): 489–526.

29 *Affiches de Paris*, 21 April 1749, n.p. (occult physics); *Avant-Coureur*, 13 March 1769, 164 (horoscopes).
30 *Avant-Coureur*, 29 October 1764, 691.
31 *Affiches de Paris*, 16 December 1792, 5182.
32 Cissie Fairchilds, "The Production and Marketing of Populuxe Goods in Eighteenth-Century Paris," in *Consumption and the World of Goods*, eds. John Brewer and Roy Porter (London: Routledge, 1993): 228–48. See Colin Jones, "The Great Chain of Buying: Medical Advertisement, the Bourgeois Public Sphere and the Origins of the French Revolution," *American Historical Review* 101 (1996): 13–40; Daniel Roche, *A History of Everyday Things: The Birth of Consumption in France, 1600–1800*, trans. Brian Pearce (Cambridge: Cambridge University Press, 2000).
33 *Affiches de province* 19 February 1784, 105–6.
34 *Journal de Paris*, 7 July 1787, 825.
35 Emile Campardon, *Madame de Pompadour et la cour de Louis XV* (Paris: Henri Plon, 1867), 408–11. Guyot, *Nouvelles récréations physiques et mathématiques*, 4 vols. (Paris: Gueffier, 1769–1770).
36 *Affiches de Trois-Evêches*, 19 February 1789, 57.
37 *Affiches de province*, 20 November 1784, 3059.
38 On Bianchi's breast pump see, for example, *Journal de Nancy*, 12:17 (1783): 44; and *Bibliothèque physico-économique, instructive et amusant* 6:1 (1787): 331. On the response of the Faculté de Médecine see e.g. *Bibliothèque physico-économique, instructive et amusant* 2 (1783), 271–3; and *Journal de Physique* 27 (1785), 198–203.
39 *Journal de Paris*, 17 February 1780, 202.
40 *Affiches de Paris*, 6 January 1746, 5–6; and ibid; 13 January 1746, 4–5.
41 *Affiches de Paris*, 19 April 1782, 922 (Bianchi); *Affiches de province*, 29 May 1784, 313 (Bienvenu's subscription package); *Journal de Paris*, 25 March 1786, 338–9 (Bienvenu on gases); *Journal de Paris*, 6 December 1786, 1416 (Famin).
42 The Paris police established, by law, the prices for the theaters. See Robert Isherwood, *Farce and Fantasy* (Oxford: Oxford University Press, 1986), 167. It is unclear, however, if popular science shows were obligated to uphold those prices or if they did so by choice. For Val see *Affiches de Paris*, 25 April 1793, 1764; and for Perrin see *Journal de Paris*, 5 April 1789, 434.
43 Pelletier, *Nouveau cabinet du sieur Pelletier* (Paris: Thiboust, 1784); and *Affiches de Paris*, 3 January 1760, 6 (group discounts).
44 *Affiches de Paris* 18 November 1777 (supplement), 1418–19.
45 *Journal de Paris*, 12 December 1781, 1391–2.
46 This book is reviewed, and its female authorship discussed, in *Affiches de province*, 24 February 1779, 31–2. It is possible that this review is referring to the work written by Anne d'Aubourg Miremont, *Traité de l'éducation des femmes, et Cours complet d'instruction*, 7 vols. (Paris: Ph.-D. Pierres, 1779–1789).
47 *Réimpression de l'Ancien Moniteur*, vol. 3, 12 March 1790, 586.
48 On the idea of a geography of science, see Steven Shapin and Simon Schaffer, *Leviathan and the Air-Pump: Hobbes, Boyle, and the Experimental Life* (Princeton: Princeton University Press, 1985), chp. 8.
49 On Nollet and, more generally, experimental physics and the educational system in Paris, see Jean Torlais, "La physique expérimentale," in *Enseignement et diffusion des sciences en France au XVIIIe siècle*, ed. René Taton (Paris: Hermann, 1964): 619–45.
50 Nollet, *Leçons de physique expérimentale*.
51 *Affiches de Paris*, 24 November 1746, n.p.
52 [Devilliers,] *Les numéros parisiens, ouvrage utile et nécessaire aux voyageurs à Paris* (Paris: L'Imprimerie de la Vérité, 1788), 86.
53 On "grand Thomas" and dentists see Colin Jones, "Pulling Teeth in Eighteenth-Century Paris," *Past and Present* 166 (2000): 100–46; on activity on the Boulevard du Temple and the

Pont-Neuf see Isherwood, *Farce and Fantasy*, *passim*; and Karen Newman, "Toward a Topographic Imaginary: Early Modern Paris," in *Historicism, Psychoanalysis, and Early Modern Culture*, eds. Carla Mazzio and Douglas Trevor (London: Routledge, 2000): 59–81.

54 On social elites mixing with the popular classes see Isherwood, *Farce and Fantasy*; and Laura Mason, *Singing the French Revolution: Popular Culture and Politics, 1787–1799* (Ithaca: Cornell University Press, 1996).

55 On wax museums see Jean Adhémar, "Les musées de cire en France, Curtius, le 'Banquet Royal', les Têtes coupées," *Gazette des beaux-arts* 92 (1978), 205–10. On the boulevard theaters see Michele Root-Bernstein, *Boulevard Theater and Revolution in Eighteenth-Century Paris* (Ann Arbor, MI: UMI Research Press, 1984), esp. 41–75. On Zaller see *Affiches de Paris*, 17 May 1770, 491; *Almanach forain*, 1773, n.p.; and *Affiches de Paris,* 23 June 1774, 559. For Noël and Val see, e.g., *Affiches de Paris*, 30 August 1779, 1934; *Journal de Paris*, 1 June 1783, 637; and *Affiches de Paris*, 27 November 1779.

56 On boulevard theater audiences see John Lough, *Paris Theatre Audiences in the Seventeenth & Eighteenth Centuries* (London: Oxford University Press, 1957), 167–8. On Perrin see *Chronique de Paris*, 1 April 1790, 364; Perrin, *Amusemens de physique* (Paris: P. de Lormel, 1787b); Perrin, *Amusemens physiques* (Paris: P. de Lormel, 1789); and Perrin, *Prospectus* (Paris: P. de Lormel, 1786). On the culture of the theater more generally see Jeffrey Ravel, *The Contested Parterre: Public Theatre and French Political Culture, 1680–1791* (Ithaca: Cornell University Press, 1999).

57 For the Palais Royal as the capital of Paris see Louis Mercier, *Panorama of Paris* (1999), 202. For general descriptions of the Palais Royal see e.g. *Petit journal du Palais Royal, ou Affiches, annonces et avis divers*, 15 September 1789, 26–8; ibid., 2 October 1789, 15; François-Marie Mayeur de Saint-Paul, *Tableau du nouveau Palais Royal*, 2 vols. (London, 1788), II:17–21.

58 Luc-Vincent Thiéry, *Almanach du voyageur à Paris*, 5 vols. (Paris: Hardouin, 1783–1787). For descriptions of the area around the Palais Royal see, for example, the volume for 1786, 324–5, 332–4, 351–2.

59 *L'Année littéraire*, 1754, IV:350–3. On his career in London see Alan Q. Morton, "Concepts of Power: Natural Philosophy and the Uses of Machines in Mid-Eighteenth-Century London," *British Journal for the History of Science* 28 (1995): 63–78, esp. 67–68.

60 On La Pouyade see *Journal encyclopédique*, May 1766, 4. See La Perrière de Roiffé, two titles, *Extrait du nouveau système générale de physique et d'astronomie* (Paris: Debure, 1761), and *Méchanismes de l'électricité et de l'univers*, 2 vols. (Paris: Paul-Denis Brocas, 1756). On the reception of these theories see *Avant-coureur*, 11 March 1766, 173–4; and *Affiches de Paris*, 23 January 1777, 107.

61 *Affiches de Paris*, 29 November 1788; and *Journal de Paris*, 6 April 1789, 437.

62 For his cabinet near the rue de Richelieu see *Avis divers: Supplement de Annonces, affiches et avis divers (Paris, 1751–1782),* 28 April 1778, 519–21; on his cabinet on the Boulevard du Temple see *Journal de Paris*, 9 May 1778, 515; and for his connections with elite science see *Avis divers*, 14 February 1778, 198.

63 On Marat's scientific work see Charles Gillispie, *Science and Polity in France at the End of the Old Regime* (Princeton: Princeton University Press, 1980), 290–330; Clifford Conner, *Jean Paul Marat: Scientist and Revolutionary* (New Jersey: Humanities Press, 1997); and *Marat homme de science?*, eds. Jean Bernard, Jean-François Lemaire and Jean-Pierre Poirier (Paris: Les Empêcheurs de Penser en Rond, 1993). Conner claims that "Marat was neither charlatan, nor quack, nor pseudo-scientist. In the context of the late eighteenth century, Marat's science was normal science." Conner, 9. Marat's most famous attack on high science, where he defines physicists as "little amateurs with grand pretensions," is "Les charlatans modernes," in *Les Pamphlets de Marat*, ed. Charles Vellay (Paris: Charpentier & Fasquelle, 1911), 255–96.

64 Jacques-Philippe Ledru, "Lettre de M. Le Dru, fils, Sur quelques Expériences de M. Marat," *Journal de physique* 18 (1781): 402–4. Jacques-Philippe Ledru was the son of Nicolas-Philippe Ledru, with whom he collaborated in the Hospice Medico-Electrique. On Bertholon see

H. Duval, "Marat et l'abbé Bertholon (1785)," *Revue historique de la Révolution française* 3 (1912): 461–2; and Charles Vellay, "Marat et l'abbé Bertholon," *Revue historique de la Révolution française* 3 (1912): 294–7.

65 Every biography of Marat contains a version of this story but the versions are not consistent. See Olivier Coquard, *Jean-Paul Marat* (Paris: Fayard, 1993), 151–5. Most accounts draw on two letters: Marat to Charles, 1783; and Marat to Lenoir, 1783, both in Marat, *Le correspondance de Marat*, ed. Charles Vellay (Paris: Charpentier and Fasquell, 1908), 14–16.

66 Charles did not even begin studying experimental physics until he was thirty years old: Bibliothèque de l'Institut (BI), ms. 2104, the Préface to "Discours d'introduction à un Cours de Physique," n.p. For biographical information on Charles see Jean-Baptiste-Joseph le Baron Fourier, "Eloge historique de M. Charles," *Mémoires de l'Académie Royale des Sciences* 8 (1829): lxxiii–lxxxviii; Anatole France, *L'Elvire de Lamartine: Notes sur M. & Mme. Charles* (Paris: Champion, 1893); and the anonymous biography in the Fonds Joseph Bertrand at the BI, ms. 2038, ff. 134–52. Charles, "Discours d'introduction à mon premier Cours de Physique lû le 22 Janvier 1781," BI, ms. 2104, pièce 1, n.p. (on the spectacle of science). On his work doing inventories see the Archives Nationale, F17–1265, f. 5; and F17–1219. On Charles's instrument collection see Fourier, xxvi. For an inventory of Charles's cabinet see AN, C-144, f. 173. On Volta and Franklin see France, *L'Elvire de Lamartine*, 19. On his election to the Institut see Maurice Crosland, *Science Under Control: The French Academy of Sciences, 1795–1914* (Cambridge: Cambridge University Press, 1992), 146. For guidebooks see Jacques-Antoine Dulaure, *Nouvelles descriptions des curiosités de Paris*, 2 vols. (Paris: Lejay, 1785), I:186; and Luc-Vincent Thiéry, *Almanach du voyageur à Paris*, 5 vols. (Paris: Hardouin, 1783–1787), II:606–8. For his courses see, e.g., *Journal de Paris*, 25 December 1780, 1466 (physics); and *Affiches de Paris*, 18 May 1786, 1302 (electricity); and M. de la Métherie, "Précis de quelques expériences électriques, Faites par M. Charles, Professeur de Physique," *Journal de physique* 30 (1787): 433–6. A brief description of all sixty lectures can be found in "Cours de Physique. Par Charles, de l'Institut National, an 11, 1802," BI, ms. 2104, pièce 18.

67 On Bouthoux de Lorget see Emile Campardon, *Les spectacles de la foire*, 2 vols. (Paris: Berger-Levrault, 1877), I:177; *Journal de Paris,* 5 April 1789, 434; and ibid., 12 April 1789, 466. On Greppin and Billiaux see *Journal de Paris*, 27 December 1781, 1454; and *Nouvelles de la république des lettres et des arts*, 22 May 1782, 151.

68 For his first course see *Journal de Paris*, 24 December 1785, 1481. For his views on physics see Famin, *Cours abrégé de physique expérimentale, à la portée de tout le monde* (Paris: Briand, 1791), i–ii.

69 On Bianchi's early career see Oliver Hochadel, *Öffentliche Wissenschaft: Elektrizität in der deutschen Aufklärung* (Göttingen: Wallstein, 2003), 187–92. On electricity see *Affiches de Paris*, 16 December 1782, 2907; on microscopes see *Journal de Paris*, 19 April 1782, 433–4. For elite savants see *Journal de Paris*, 26 August 1782, 972–3.

70 For demonstrations of gases see *Affiches de province*, 19 February 1784, 105–6; for his class in electricity see *Journal de Paris*, 1 May 1786, 491; on light *Journal de Paris*, 29 August 1787, 2052. Launoy and Bienvenu, *Instruction sur la nouvelle Machine inventée par MM. Launoy, Naturaliste, & Bienvenu, Machiniste-Physicien* (N.p., n.d.).

71 For the "epidemic curiosity" see *Avant-coureur*, 9 July 1759, 346; on Ledru's experiments see ibid., 8 April 1765, 210–11; and for his combination of reason and entertainment see *Almanach forain*, 1778, 132.

72 On Ledru's demonstrations before the king see [Roze de Chantoiseau,] *Essai sur l'Almanach général d'indication d'adresse personnelle et domicile fixe, des six corps, arts et métiers* (Paris: La veuve Duchesne, 1769), n.p. For a review of Ledru's work see, e.g., *Journal de physique*, 1775, 195–6; for Ledru's license see AN, O1–126, f. 407; and O1–590, f. 1; and for his treatment of incurables see Louis Petit de Bauchaumont, *Mémoires secrets*, 30 September 1783. Ledru published the report issued by the academicians, along with an introduction outlining his ideas on electricity and its applications. Ledru, *Rapport de MM. Cosnier, Maloet, Darcet, Philip,*

Le Preux, Desessartz, & Paulet (Paris: Philippe-Denys Pierres, 1783). For use of the title "King's physicist" see *Almanach forain*, 1786, 154; on the approbation of the Faculté de Médecine see Luc-Vincent Thiéry, *Guide des amateurs*, 2 vols. (Paris: Hardouin & Gattey, 1786–1787), II:688–9. On the Angora cat see *Journal encyclopédique*, June 1784, 310–11.

73 See Geoffrey V. Sutton, "Electric Medicine and Mesmerism," *Isis* 72 (1981): 375–92; Jean Torlais, "Un prestidigitateur célèbre, chef de service d'électrothérapie au XVIIIe siècle, Ledru dit Comus (1731–1807)," *Histoire de la médecine* 5 (1955): 13–25. Ledru still operated his Hospice Médico-Electrique in 1792: see *Affiches de Paris*, 1 October 1792, 4143.

74 On Pelletier's early career see Thiéry, *Guide des amateurs*, I:495. On the fire at the Saint-Germain fair see Cherrière, "La lutte contre l'incendie dans les Halles, les marchés et les foires de Paris sous l'ancien régime," *Mémoires et documents pour servir à l'histoire du commerce et de l'industrie en France* 3 (1913), 252–63. On the various locations of Pelletier's shops see *Affiches de Paris*, 10 may 1759, 295; and *Avant-Coureur*, 12 October 1761, 649–50; ibid., 13 June 1763, 377. *Correspondance littéraire*, 16 vols., ed. Maurice Tourneux (Paris: Garnier, 1877–1882), VIII:444; Bauchoumont, *Mémoires secrets*, 12 June 1783. On his travels see *Affiches de Bordeaux*, 20 May 1773, 88; and *Affiches de l'Orléannais*, 4 June 1779, 95. For a glimpse into some of his troubles during the French Revolution see Pelletier, *Exposé succinct des torts du sieur Alexandre Barré, ancien garçon boucher* (1790).

75 The first mention I have of Pelletier using the appellation "king's engineer and machinist" comes from the *Affiches de l'Orléannais*, 4 June 1779, 95. Pelletier first mentions his royal patron in 1781, but fails to mention how he managed to obtain it; see *Affiches de province*, 13 March 1781, 43–44. Pelletier, *Hommage aux amateurs des arts* (Paris: Thiboust, 1782).

76 For his mirror demonstration see *Nouvelles de la république des lettres et des arts*, 31 July 1782, 231–2. On his short courses see *Journal de Paris*, 23 December 1785, 1477; and Thiéry, *Guide des amateurs*, II:495.

77 AN, O1–1293, f. 90 (for Pelletier's request for a pension); *Affiches de Paris*, 4 June 1790, 1574 (on his club). For his pension during the French Revolution see *Archives parlementaires de 1787 à 1860*, series I, LXXIII:210; and *Procès-verbaux du Comité d'instruction publique de la Convention nationale*, ed. M.J. Guillaume, 6 vols. (Paris: Imprimerie Nationale, 1891–1907), VI:362–3.

4

Institutions of popular science

Eighteenth-century Parisians found themselves inundated with a plethora of new clubs all vying for their attention, patronage, and financial support.[1] Many of these organizations attracted members by linking themselves explicitly with common Enlightenment themes and by claiming to disseminate useful knowledge.[2] These societies often acted as centers for the transmission of enlightened ideas to the general public. In the last decades of the eighteenth century, one type of club, usually called *musées* or *lycées* after mythological muses or ancient Greek schools, particularly captured the imagination of many Parisians, and news of their activities and meetings filled the popular press. As one proponent of *musées* claimed, "it is the idea of offering, as it were, to each mind a nourishment which is appropriate to it, for encouraging every taste, and for obtaining an open worship for the sciences and the arts, that has given birth to the project of the *musée*, consecrated in the name of public utility."[3] That usefulness frequently came in the field of the sciences, and the *musées* often found themselves associated with the scientific side of the Enlightenment.

Incorporating a mixture of philosophical salon, scientific academy, reading society, and public lecture, *musées* quickly entered the late eighteenth-century enlightened cultural milieu and won the approval of the public. *Musées* provided their members, for an annual fee, access to the world of science, literature, and ideas. Usually organized by a single individual, the *musées* provided a private space, away from state control, for a variety of activities; these ranged from guest lecturers speaking about their new books to permanent teachers offering regular courses on a wide array of topics. Libraries, scientific laboratories, and salons were standard parts of the *musée* experience. In the years before the French Revolution, more than ten such clubs appeared, or were under consideration, in Paris. In these clubs a loosely defined general public could meet to discuss the day's events, read the latest journals or books, experiment with scientific machines and instruments, debate philosophy, or, most commonly, attend a series of lectures on a particular topic. Lecture courses, usually based on similar public courses already offered throughout Paris, provided the main focus of events in the *musées*. These clubs drew their members from a fairly broad spectrum of men and women ranging from the nobility to well-off artisans. Membership required only the sponsorship of individuals who were already members and the payment of an annual subscription fee that ranged from as little as fifteen livres to as much as several hundred livres.[4] Once a member, however, individuals usually enjoyed all the benefits of the *musée* regardless of gender, social origins, or economic standing.

Musée organizers offered an increasingly dynamic and interested Parisian public the chance to participate in the Republic of Letters, rather than just reading about it, by providing a formalized system of public education as well as a social space where subscribers could interact with savants and other amateurs. Active in most of the major French cities, the founders of *musées* consciously carved out an important role in the spread of enlightened ideas, particularly within the realm of science.[5]

Popular science and the Republic of Letters

Studies of Enlightenment culture from the last quarter century have made explicit reference to the importance of the *musées*. Dena Goodman, for example, analyzed them in conjunction with salons in *The Republic of Letters: A Cultural History of the French Enlightenment*.[6] Goodman argued for the dominance of salons within the culture of the Enlightenment. *Musées*, she claims, were largely male-dominated clubs that denied women the opportunity to act in a governing role as they had in the salons.[7] Another historian, Hervé Guénot, has concentrated on the *musées* as a space for sociability.[8] Unlike Goodman and Guénot, most analyses of the *musées* have been undertaken by historians of science who usually emphasize them as places where elite savants diffused science to the willing, but uneducated, general public.[9] Other, older, discussions of these clubs concentrate on their place as education institutions or on their connections with freemason lodges, a link established largely through similar membership.[10] *Musées*, however, should be placed within the larger context of the emerging public sphere and popular science. Rather than focusing only on their relationship with freemason lodges or the participation of elite savants, the importance of *musées* lies with their fusion of various types of Enlightenment institutions and as forums for the dissemination and appropriation of knowledge.

Musées provided Parisians with a social space where they could gather to listen, read, and discuss ideas. They were one of a number of such spaces, with various styles, formats, and purposes, available all over France.[11] In Paris, salons had long held a place of honor as sites of enlightened sociability, while cafés drew an eclectic audience often interested in discussing the day's news or the latest ideas and philosophical trends. Alternatively, *cabinets de lecture*, established beginning in the 1770s, allowed members to read books and journals without having to buy them. Masonic lodges also supplied a social space, in this case secret, where men and women met to discuss and disseminate enlightened ideas. Provincial academies fulfilled similar roles outside of Paris, but new academies were not allowed near the major institutions already established in the capital.[12] The *musées* occupied a key place in this profusion of Enlightenment institutions. It was not enough, noted Louis-Sébastien Mercier, to read some new pamphlets, meet some philosophers, or read some literature: "we must still have the *lycées*."[13] Less exclusive than academies, salons, or freemason lodges but more formal than cafés or reading societies, the *musées* established a new type of club, one that formalized the system of public lecture courses that had been growing throughout the eighteenth century. Although *musées* incorporated organizational ideas from salons, lodges, academies, and cafés, the key to their success lay in the opportunities they afforded for amateurs to come into contact with enlightened culture through the centralization of multiple lecture courses under one roof.

Science played an important part in the institutionalization of a popular Enlightenment, although certainly not the only part, and the *musées* contributed to the creation of scientific culture in this environment. Specifically, the individuals who ran the *musées* consciously sought to act as intermediaries between science and the general public by supplying the appropriate materials of Enlightenment – books, laboratory equipment, and forums for the discussion of ideas – and establishing a specific location where men of letters and interested amateurs could meet and interact in a manner designed to be mutually beneficial to both. *Musée* members gained access to the Republic of Letters in all its many manifestations as well as the opportunity to attend classes given by respected thinkers. For their part, the popularizers gained an audience for their work, a chance to disseminate their views on a variety of subjects, and the social legitimacy of popular backing along with the economic rewards of teaching to a more general audience. The latter two gains were especially important in a time of limited state support and competitive financial patronage. In this respect, the *musée* organizers, like the purveyors of public lecture courses, closely resemble the individuals described by Robert Darnton as "grub street philosophes."[14] Unable to acquire official patronage or positions, they created their own philosophical space and bided their time, waiting for more traditional types of support to make itself available.

The broad range of learning available at the *musées* and their avowed goal of making that information useful and accessible to members gave the *musées* a significant role in the encyclopedic strand of the Enlightenment which sought to codify, define, and disseminate all human knowledge. The encyclopedic side of the Enlightenment continually sought to make knowledge available and comprehensible to as many individuals as possible, and public experimental physics courses, with their emphasis on visually reproducible knowledge, played a key role in that endeavor.[15] In many ways, this popularized version of the Enlightenment probably reached a greater portion of the general public than did its other, more philosophical aspects. For their part, the *musées* provided a specific location for the appropriation of science and a formal and centralized site where all intellectual genres were made accessible for the general benefit of the members.

This traffic in news and exchange of ideas attracted many people either seeking to learn something of value or, sometimes, just looking to occupy their leisure time in an entertaining but simultaneously useful fashion. Located firmly in the public sphere, *musées* contributed to the diffusion of enlightened science by creating a center for the exchange of ideas. Scientific culture, within the public sphere, had come to be something of a commodity.[16] It was bought, sold, and traded in the form of popular periodicals, subscription fees to reading societies, or the price of the cup of coffee you had to buy to join in the conversation or read the newspapers at your local café. Scientific cultural capital could take several forms including the purchase of material objects, such as books, scientific instruments, or natural history artifacts. It could also, on the other hand, be less tangible and come in the form of knowledge acquisition.[17] The commercial trade in ideas held a significant place in eighteenth-century Paris as knowledge of scientific language, methods, goals, and discoveries became a necessary and significant part of contemporary culture. In all of this the *musées* occupied a key role as one of the new locations for the consumption of knowledge.

Musée membership grew to encompass a large and varied audience. Although complete subscription lists are not available, we know that the largest one, the Musée de Monsieur, had at least seven hundred members, more than ten percent of which were women, and so was larger than any comparable salon or academy. Another, the Salon de la Correspondance, boasted a membership of more than three hundred people and a subscription list for its journal exceeding five hundred. Overall membership for *musées* in late eighteenth-century Paris probably reached as high as several thousand individuals. But how did these clubs come to occupy such an important place amid the growing number of eighteenth-century social and cultural institutions?

The origins of scientific clubs in late eighteenth-century Paris

Musées gained notoriety by fusing the best aspects of other institutions together in one social space and then adding lecture courses into the resulting mixture. While largely similar in outlook and intent, *musées* do show a variety of organizational structures and social set-ups. Some resembled salons in form and in the ways their members interacted. Others appeared to be miniature scientific academies, albeit different from their royally-established brethren in the nature of the audience they served, the number of members who could join, and the freedom with which members could be chosen. The clearest inspiration for a majority of *musées*, however, came from public lecture courses with their interest in useful and entertaining knowledge made available to a socially and economically eclectic audience. *Musée* organizers gathered those courses, previously scattered throughout Paris, and pulled them together under one roof.

The first club of this kind was the Salon de la Correspondance, sometimes referred to as the Correspondance Générale, established in 1777 by Claude-Mammès Pahin de la Blancherie, an enterprising young man seeking to make his mark in the Republic of Letters.[18] Noted for his "courageous zeal," La Blancherie strove long and hard to provide links between savants and artisans from all nations within the confines of the Salon de la Correspondance.[19] He designed this club as a place where individuals in Paris could discuss new ideas and trends, ranging from art to science to literature, away from other, more restrictive institutions like academies and salons.[20] In other words, La Blancherie's club provided an alternative space for the institutionalization of the Republic of Letters.[21] He wanted his office to act as a conduit for news, information, and ideas. One of the chief objects of the Salon de la Correspondance was to "instruct savants, men of letters, artisans, and amateurs."[22] As was true for many of the *musées*, La Blancherie conceived of his project as an attempt to cover the different aspects of the Enlightenment and to disseminate useful knowledge to a broad range of people.

To accomplish these broad goals, his project included several facets.[23] First, La Blancherie edited a journal, the *Nouvelles de la République des Lettres et des Arts*, that related all of the latest news of the Enlightenment.[24] This journal, with as many as five hundred subscribers throughout Europe and North America, contained letters from members, descriptions of the activities of the club, and general news of the Republic of Letters.[25] The cost for an annual subscription was twenty-four livres for Parisians and thirty livres for anyone outside Paris.[26] Although generally well respected, the journal did not appeal to everyone. La Blancherie had persuaded John Adams, then ambassador to France, to join his club and subscribe to the journal, but Adams was less than pleased to continue receiving the journal after he returned to the United

States, especially since he usually had to pay the postage. Adams begged his replacement, Thomas Jefferson, to negotiate an agreement with La Blancherie to stop sending the journal.[27] In addition to editing the journal, La Blancherie also acted as an agent for mail, receiving, forwarding, and sending letters between individual members and, when appropriate, publishing those letters in his journal. He had actually started his enterprise with the idea that the journal and the correspondence reproduced therein would be the main focus. Instead, he found that more and more people stopped by his office to talk with him and to interact with other members.

By 1779, he began operating the formal Salon de la Correspondance, sometimes called the Assemblée Ordinaire, which met on a weekly basis and whose activities were reported in the journal. Science played a very conspicuous role during these meetings.[28] Men would gather on Wednesday mornings in the Bureau de Correspondance (La Blancherie's main office) to view and discuss the objects contributed by different artists, scholars, and savants. The next issue of the *Nouvelles de la République des Lettres et des Arts* would contain a description of these exhibits, as well as a synopsis of the conversation and commentary about each piece. In April 1779, for example, the assembled group examined and discussed, among other things, a bas-relief by the sculptor Rosset de St. Claude; a bust of Parmentier, author of works on chemistry and animal economy; several flowers and plants from India; and a new kind of thermometer presented by a M. Bourbon. Science and technology found regular expression at these weekly meetings with a particular emphasis on their utility. In May 1779 Guyot, the author of the popular science book *Nouvelles récréations physiques et mathématiques*, attended the assembly and demonstrated a machine of his own invention that would aid in combating fires. Other items of scientific interest displayed at the assembly included a microscope created by the optician Dellebarre and approved by the Académie Royale des Sciences; a new electrical machine constructed by Girardin, a manufacturer of scientific instruments who also offered popular science courses; and a weather prediction device called the "Forecaster," developed by an unnamed savant but available for purchase at Jacques Bianchi's scientific instrument shop.[29]

Partially due to its attention to science, Pahin de la Blancherie's assembly earned the endorsement of the Académie Royale des Sciences. The academy had appointed a commission to look into this new club. Composed of Benjamin Franklin, Joseph Jérôme de Lalande, and the marquis de Condorcet, the commission concluded that the Salon de la Correspondance was "very useful to the progress of the sciences and the arts."[30] These academicians were particularly impressed by the range of individuals, both from Paris and from throughout Europe, who participated in the club's activities. La Blancherie's work, they suggested, actually sped up the progress of human knowledge, and, as a consequence, he and his project should receive active encouragement from the academy. This approbation held greater significance than just peer support, however. In order for an institution to survive in Paris it had to have state approval. A club like La Blancherie's, with its explicit focus on scientific matters, walked on dangerous ground: if the members of the Académie Royale des Sciences, as representatives of the state, decided that it threatened their own work in any way, the Salon de la Correspondance would have been shut down. The sanction of the academicians, then, spoke volumes concerning the place of the La Blancherie's club specifically, and the role of *musées* in general, in Parisian society. Although loosely based on royal academies, they were not trying to mimic those institutions and so in no way threatened the work of their members. Rather, La Blancherie attempted to act as a mediator between the academies and the general public. To do this, he drew on the authority of scientific work and coupled it with the willingness of amateurs to spend their leisure time and money on learning about science. Like some public lecture courses, those that lasted just one day, La Blancherie displayed certain scientific instruments and inventions, but he did not work actively to create scientific ideas or theories. So the Salon de la Correspondance was not seen as a rival to the academies since it did not produce new science, but acted instead as a medium for the dissemination and consumption of science.

Most contemporaries acknowledged that the Correspondance Générale was the foundation on which all "*lycées*, *musées*, clubs, and other establishments of that type" were based.[31] Not that everything went well for La Blancherie. He was perpetually short of money and continually sought new subscribers to solidify the club's financial basis. In 1781 he even tried to bait new members with a small lottery, hoping the greed of potential subscribers would lure them into joining.[32] Official approval and general accolades were not enough to guarantee the survival of the Salon de la Correspondance. Raising membership became an even more difficult task with the advent of other *musées* over the course of the 1780s. La Blancherie grew jealous of his rivals' success. At one point, when he had to shut down temporarily, he lost many members to the Musée de Monsieur. Despite the approbation he had received from the Académie Royale des Sciences, some of his enemies even suggested that the Musée de Monsieur should absorb the Salon de la Correspondance and take over all of its functions. The editors of the *Mémoires secrets* virulently attacked both La Blancherie and his club. They criticized the Académie Royale des Sciences for its actions and compared La Blancherie to a charlatan.[33] Others castigated La Blancherie's journal, calling it a "pitiable rhapsody of vague ideas and common anecdotes, without substance, without interest, and in a most bourgeois style."[34] Despite this vehement disapproval, La Blancherie weathered the vagaries of public opinion and continued to attract many members.

Membership in the Salon de la Correspondance came from all levels of the Republic of Letters and society in general. There were three levels of membership. At the top were the protectors – noblemen and the social elite who held a premier place in the club, and who paid considerably for that privilege. The ninety-four protectors in 1782, for example, each contributed ninety-six livres annually to maintain their membership.[35] Among these were the counts of Provence and Artois (the brothers of King Louis XVI), the dukes of Montmorency, Tonnerre, Fleury, and Polignac, the bishops of Limoges and Senlis, and other counts, viscounts, and lesser noblemen.[36] Next to the protectors were the associates and subscribers who paid forty-eight and twenty-four livres per year respectively. In 1782 there were fifty-six associates and one hundred sixty-three subscribers, by far the largest single class. Among the associates could be found bankers, businessmen, teachers, engineers, authors, clergymen, and artists. The subscriber class, claimed La Blancherie, was composed largely of artisans. One observer wrote that at the Correspondance Générale "great nobles mixed with scholars and artists: a geometer, a lockmaker, a painter, a musician, etc."[37] La Blancherie wanted to organize a socially eclectic group to whom Enlightenment science would be of most use.

La Blancherie could not boast about any particularly well-known (from a modern perspective) savants in his club. With only a few exceptions, the savants who did join were middling savants who displayed their machines, books, and instruments in the general assembly. While these are not remembered among the major figures of the Enlightenment, their contemporaries held them in relatively high esteem. These individuals include the popularizer Sigaud de la Fond and Antoine Court de Gébelin, who actually went on to found his own *musée*. The lack of female membership formed another obvious absence. Although noblewomen could pay their ninety-six livres and become protectors, La Blancherie did not admit women to the Assemblée Ordinaire.[38] "Ladies of high consideration," however, were allowed to "satisfy their curiosity" and, after the fact, examine the objects that had occupied the general assembly. This exclusion does not represent the wishes of La Blancherie, who would have welcomed interested women into his club, provided they could pay the membership fee. The objects on display, he noted, were as "equally useful and interesting to women" as to men.[39] It is probable that the state, in defining the Salon de la Correspondance as a group similar to an academy but not producing new science, would have frowned on female members just as women normally were not allowed access to the meetings of the Académie Royale des Sciences. It is equally likely that men might have preferred the exclusion of women; this had the dual result of making the meetings seem more like the academies and less like salons, while at the same time it prevented women from being in a position to direct the intellectual work of the *musées* as they did in salons.

Aside from the lack of female leadership and influence, the format of this club most closely resembled that of the salons.[40] But instead of *salonnières* guiding the ebb and flow of the conversation, individuals gathered around a specific object that then provided the impetus for the ensuing discussion. These attempts at disseminating science, then, were actually aimed at a relatively educated audience, one that could readily appreciate the implications of the objects they examined. Participation required both some knowledge about science, or at least an ability to speak generally about things scientific, and an awareness of the rules for polite conversation. In fact, for some individuals, the Salon de la Correspondance operated too much like a salon since it tended to be dominated by rules of politeness and sociability and not necessarily by scientific skill or merit. They preferred the more formal structure of the Académie Royale des Sciences and it was not long before someone established a *musée* that followed that model more strictly.

The Musée de Paris and other clubs

While the Salon de la Correspondance tried to integrate the salon with the academy, the Musée de Paris attempted to turn a private academy into a salon. This began in 1780 when Antoine Court de Gébelin founded the Société Apollonienne, rechristened the following year as the Musée de Paris.[41] Court de Gébelin clearly utilized the format of the royal academies as his guide in establishing the Musée de Paris. He wanted to create a space where savants could communicate their ideas to each other and to the general public. As stated in the official memoirs: "If the union of men into a society was necessary in order to give birth to savants, then the union of savants and the reciprocal communication of their Enlightenment was more necessary for the progress of the sciences."[42] The useful work of the savants, thought Court de Gébelin, must be made known to society for the glory of France and for the admiration of foreigners. The hope was that the Musée de Paris, "by protecting itself from the epidemic of *bel esprit*, from scientific jargon, [and] a ranting and pompously soporific style," would merit the approval of the nation and of the enlightened public.[43] The Musée de Paris wanted to become a place where "young authors tried out their talents, [and] where the best known authors presented, for the taste of the public, a manuscript that they were going to have published."[44] In creating his *musée*, Court de Gébelin specifically formed a space below the academies, but above the informal structure of public lecture courses.

Court de Gébelin was a fairly well known author whose standing within the Republic of Letters aided the success of his *musée*, although his predilection for Mesmer's animal magnetism did discredit him in the eyes of some contemporaries.[45] Nonetheless, he worked diligently to forge a place for his *musée*, described by one contemporary as the "*musée* par excellence," within the urban cultural milieu by providing a central location where men could meet to discuss new ideas, read books, or attend public lectures.[46] Not that Court de Gébelin and his *musée* did not suffer occasional reproach. One critic, for example, referred to the Musée de Paris as "an imperfect shadow of the academies."[47] Even worse for its reputation, an internal schism formed over the direction of the Musée de Paris and Court de Gébelin found himself confronted with rivals who wanted to take control of his institution.[48] The ensuing battle weakened the club and ultimately prompted many of its members to flee elsewhere, especially to the Musée de Monsieur. The infighting also led to personal attacks against Court de Gébelin, and it is at this point that his embrace of animal magnetism came under direct attack.[49] Brissot claimed that this battle ruined Court de Gébelin, adding that "hell is no worse than a *musée*."[50]

Nonetheless, before the division, Court de Gébelin worked hard to create a forum where savants could meet and share ideas. Membership was open to savants, artisans, and amateurs alike and included such savants as Bernard-Germain-Etienne Lacépede and Alessandro Volta. The *musée* boasted numerous foreign correspondents, including some from the New World. Benjamin Franklin, for example, was honored at the Musée de Paris in 1783 for his role in spreading republican ideas in the newly founded United States.[51] Although exact membership figures are not available, the meeting hall for the *musée*, with seven hundred seats, suggests a very substantial size. Anyone could attend the monthly public meetings, including ladies and foreigners.[52] However, as with the Salon de la Correspondance, while women could become honorary associates, most other activities were reserved exclusively for men.[53]

The reading of scholarly memoirs on a variety of subjects formed the chief activity of the Musée de Paris. "We have come to establish here," wrote Court de Gébelin, "a private society of sciences, letters, and arts. We propose to give memoirs on the state of the sciences, letters, and arts in Europe."[54] These presentations would then be published in the annual *Mémoires du Musée de Paris*, in obvious imitation of the *Mémoires of the Académie Royale des Sciences*. With this journal, Court de Gébelin intended to disseminate new scientific ideas as quickly and broadly as possible for the benefit of both savants and the general public. He emphasized the speed of publication since the Académie Royale des Sciences memoirs often appeared after a considerable delay, making it difficult for academicians relying on that source to claim precedence for their discoveries or ideas. To give an example, one issue of the memoirs included a poem praising Pilâtre de Rozier and the marquis d'Arlandes for making the first voyage in a hot air balloon, an ode to electricity, and a lengthy treatise on the nature of fire.

Court de Gébelin's *musée* retained certain well-known savants to offer courses and give lectures. Again, science, in all of its facets, contributed significantly to this program. One member, writing in the *Mémoires of the Musée de Paris*, claimed that the *musée* would admit any scientific work that presented a good argument based on facts or experiments. He noted that refusing to examine controversial works could be regarded as destructive to progress and added that even erroneous ideas had sometimes led to useful discoveries.[55] Despite its good intentions, the Musée de Paris ultimately offered few lectures.[56] After the schism formed it found itself unable to fulfill many of its promises.

Dissatisfied with these early attempts to address the needs of both savants and the general public, some later founders of *musées* formed establishments with a more specific audience or purpose in mind. Madame la Baronne Duplessy for example, established the Musée des Dames. She observed that in spite of the existence of institutions and spaces for the instruction of men, such as academies, *musées*, and universities, there did not exist any that were expressly for women.[57] In fact, as Duplessy noted, only a few institutions at this time even admitted women to their ranks.[58] In an effort to change that pattern she formed the Musée des Dames where "all the knowledge that was suitable for ladies" could be acquired, and the entertaining was joined with the useful. The lectures concerned many topics, with the sciences holding an important place. In particular, Duplessy supported classes in experimental physics, a science well suited to amuse and instruct the members as long as the disseminators avoided rehashing what she labeled the "ridiculous dispute" between attraction and vortices (i.e., Newtonianism and Cartesianism).[59] The Lycée des Femmes offered another alternative; its founders expressed the hope of instructing women on the importance and the usefulness of the sciences and the arts. The members of this *lycée* could attend, along with a multitude of other classes, a course in mathematics and experimental physics that met for two hours twice each week.[60] Many other *musées*, like the Lycée des Arts, were founded during the revolutionary period.[61] Nonetheless, although many of the other *musées* were quite popular, the Musée de Monsieur held the premier place among these clubs.

The Musée de Monsieur

Both La Blancherie and Court de Gébelin had already established their societies when, in 1781, the Musée de Monsieur joined the fray.[62] This *musée*, founded by Jean-François Pilâtre de Rozier, quickly gained a reputation for excellence and was soon considered by many to be "infinitely preferable" to most of its rivals.[63] The Musée de Monsieur proved to be the most popular institution for the dissemination of science thanks largely to Pilâtre de Rozier's ability to bring elite and amateur savants together with men and women from the noble and middling classes. Pilâtre de Rozier's club achieved its dominant position through a combination of good planning, useful connections in high places, and a string of very fortunate circumstances.

The history of this *musée* stretched well into the nineteenth century, making it the most durable of all such institutions founded in the decade before the Revolution, although it did go through several name and organizational changes along the way.[64] It was founded under the auspices of the count and countess of Provence, after whom it was wisely named.[65] Pilâtre de Rozier led it under this name from 1781 until his death in 1785; from 1785 to 1793 it was known simply as the Lycée, during which time a consortium of rich individuals, who had purchased the bulk of Pilâtre de Rozier's estate and donated it back to the club, operated it; during the Revolution it judiciously became the Lycée Républicain and began to eschew its noble patrons in favor of a more democratic system, becoming for a time the property of eighty-seven shareholders given the title of founders; in 1802, after the creation of educational institutions called *lycées*, it changed its name to the Athénée de Paris. In 1814, for its final incarnation, it took on a more distinctively elite character and became the Athénée Royal de Paris, a name it kept until it disbanded in 1848.[66]

The popularity of the Musée de Monsieur had a great deal to do with its founder. Pilâtre de Rozier laid the groundwork for his future achievement early in his career by working diligently to accumulate both patrons and positions.[67] After a brief time studying pharmacy in Paris, he turned his attention to experimental physics and chemistry where his efforts caught the eye of several academicians including Balthazar-Georges Sage. Sage arranged a teaching job for him in Reims at the Société Libre d'Emulation, an establishment dedicated to encouraging and promoting interest in the arts and sciences.[68] Returning to Paris in 1780, and making liberal use of his natural charms and academic connections, Pilâtre de Rozier acquired the post of intendant of the physics, chemistry, and natural history cabinets in the household of the count and countess of Provence.[69] At the same time, his research began to gain the respect of his colleagues and he slowly entered into numerous facets of scientific life. He joined state-sponsored scientific commissions, for example, and published his research on the causes of thunder and on optics in the *Journal de physique*, where it underwent general scrutiny and garnered him widespread admiration among his colleagues and the public.[70] In addition, Pilâtre de Rozier gained fame in Paris for his public lecture courses in experimental physics, where he covered popular topics like optics, magnets, and electricity.[71] He also taught classes in chemistry, focusing explicitly on the subject of gases, which was both an early interest of his and a topic much in vogue among the general population following the first balloon flights. His participation in the craze that followed the Montgolfier brothers' first experiments with hot-air balloons placed him firmly at the forefront of popular science. Pilâtre de Rozier gained international fame as the first man to ascend into the heavens in a hot-air balloon; unfortunately, he also earned eternal infamy as the first man to die as a result of falling from the sky in what was the world's first fatal ballooning accident.[72] Nonetheless, his combination of social and academic connections, along with his popular charisma, propelled Pilâtre de Rozier's bold career during which he helped make science readily available in the public arena.

Like La Blancherie and Court de Gébelin, Pilâtre de Rozier fused together several institutional traditions in creating the Musée de Monsieur. Salons, academies, and public lecture courses all acted as models for the *musée*, although in this case it was the latter style that predominated. At the same time he created something new and different. He fused the credibility and respectability of the academies, the sociability and intellectual conversation of the salons, and the entertaining but educational structure of public lectures to create an organization that brought Enlightenment ideas to an audience drawn from all three of these traditions. Members could take advantage of the permanent and public location and, for one annual fee, attend a wide variety of lecture courses in both the sciences and the humanities. The actual physical space of the Musée de Monsieur facilitated its ability to meet a variety of social, scientific, and intellectual needs.[73] The rooms contained a laboratory with all the machines and instruments necessary for the study of chemistry; two *cabinets de physique*; a library containing numerous works on the sciences as well as a number of popular and scientific periodicals; a salon where new machines and other curious works were displayed for the benefit of members; and a lecture hall where professors taught their courses.[74] Science had such an important place here that even contemporaries referred to this club as the "scientific *musée*."[75] This characterization is somewhat false since members could also attend courses in nonscientific areas including languages, literature, and history.

The Musée de Monsieur attempted to address several interrelated goals. "An institution whose essential object [is] the study of the sciences and the knowledge of the arts," it hoped to provide a space "where citizens from all levels of society might acquire some Enlightenment, and savants could enjoy a glorious satisfaction in communicating that knowledge to them. It is an establishment, where the most beautiful qualities are those of knowledge and charity" consecrated in the "public interest." More generally, it satisfied "the taste of the French people for the sciences." The entire nation, claimed Pilâtre de Rozier, would "receive the inestimable advantages" of the Musée de Monsieur.[76] Whether this *musée* actually succeeded in these aims is not the question here. Gauging reactions to the club remains difficult since only a few descriptions of its activities are extant. Nonetheless, Pilâtre de Rozier's *musée* illustrates the importance of popular scientific institutions in eighteenth-century culture.

Specifically, these advantages were threefold. First, the *musée* acted as a center for education and the dissemination of current ideas and trends in science. The intent, claimed the prospectus for 1788, was "to plant the seeds of desire for self-instruction." To this end, the Musée de Monsieur employed a wide variety of instructional methods designed to fit the individual needs of its members. Primarily, it offered a number of classes throughout the year. Courses in the sciences were available almost daily and the total number of options increased steadily. With a broad range of topics addressed by the professors, the *musée* itself strove for a nearly encyclopedic coverage of the sciences and arts. In 1788, for example, a member could attend lectures on chemistry, natural history, and botany by Fourcroy; mathematics and experimental physics by Antoine Deparcieux; theoretical physics by Gaspard Monge; anatomy and physiology by Jean-Joseph Süe; as well as literature by Jean François de la Harpe; history by Jean-François Marmontel and Dominique-Joseph Garat; and other classes in English and Spanish.[77] Many of these professors had offered versions of their courses to the general public in the years before the organization of the *musée*.

For members looking for more individualized instruction, other options were available. Two *cabinets de physique*, stocked with all the latest machines and instruments, were set up so people could perform or watch experiments.[78] Members could actually duplicate the demonstrations they had witnessed or read about and, at least theoretically, participate in the exercise of their own reason and personally witness the subjugation of nature to humankind. The ability to actually do experiments created a sense of individual involvement in the Enlightenment enterprise. If they were not interested in performing experiments, members could always read about some other savant's scientific exploits in a library containing over eleven hundred titles. Physics occupied a big place here, and works by both elite and middling savants could be found on the shelves. Isaac Newton's *Opticks*, Benjamin Franklin's *Experiments on Electricity*, and Joseph Priestley's *History and Present State of Electricity* were available, along with many of the basic handbooks of scientific popularization such as Jean-Antoine Nollet's *Leçons de physique*, Joseph Aignan Sigaud de la Fond's *Description et usage d'un cabinet de physique*, and Charles Rabiqueau's *Microscope moderne*. Both popular and scientific journals, including the *Journal de Paris*, *Affiches de Paris*, *Affiches de province*, *Journal de Physique*, and *Journal des Savants*, were also available. This varied collection of books and periodicals clearly indicated an effort to reach as wide a public as possible.

The second advantage of membership in the Musée de Monsieur centered on the practical value of that institution.[79] Emphasis fell on the utility of knowledge and the desire to impart worthwhile information, whether for savants, amateurs, or artisans, rather than on creating a space where only elites would sit and talk.[80] Pilâtre de Rozier hoped that in addition to imbuing members with a taste for the sciences, his *musée* would concretely help individuals acquire knowledge in the mechanical arts, sciences, and commerce. In theory, only the most interesting and useful aspects of eighteenth-century thought were presented for the appropriation of the public.[81] The course in mathematics, offered by Condorcet, provides a case in point. In the introductory remarks to his class he claimed that the object of the Musée de Monsieur was "not to form savants, but to give, to those men who may not have it, the possibility of using the parts of the sciences which are of the most immediate utility to them, and to make them understand simple but certain principles which will preserve them from the errors in which their proper imagination or the prestige of charlatans might make them fall." Condorcet implicitly addressed himself to a middling audience and the attempt to create a popular scientific culture that would allow amateurs to recognize both the genius of the elite savants and the failings of the scientific charlatans. Condorcet felt that the courses given at the Musée de Monsieur enabled audiences to understand the true fundamentals of science, prevented them from putting forth ridiculous ideas or mispronouncing scientific words, and allowed them to avoid being duped by fanciful systems composed of unintelligible and vague principles. The successful course, he thought, should "render reason to all."[82] Ultimately, claimed Condorcet, it was the utility of the knowledge available in the *musées* that made them an invaluable resource for Parisians.[83] This idea, that the Musée de Monsieur was consecrated to "public utility," echoed throughout most contemporary descriptions of its role in society and created a specific link between the work of the *musées* and the goals of the Enlightenment in general.[84]

The third advantage of membership in the Musée de Monsieur was access to a vast array of amusing and entertaining demonstrations and displays. Not everybody who joined this *musée* sought to improve themselves or, more precisely, not all learning needed to be dry or dull. On the contrary, from what we know of the lectures given at the Musée de Monsieur, it would seem that several of the instructors drew heavily on the pedagogical methodology prominent in the public lecture courses. Pilâtre de Rozier himself utilized overtly theatrical methods to attract and hold the attention of his audience. Most spectacular, perhaps, was his demonstration of the volatile nature of hydrogen by inhaling some and then blowing it out through a tube over a candle flame. The resulting fireball explosion immediately captured the attention and the imagination of everybody in the room. Another theatrical lecture, designed to explain a complex phenomenon for a general public, was Pilâtre de Rozier's description of Newtonian gravity. He equated gravity with love, claiming that just as men are drawn to certain women, astronomical bodies undergo gravitational attraction. "Assume, ladies," he wrote, "that I am between two of you; each is worthy to be loved, but nevertheless my heart must choose. Well, attraction might draw me more to the lady on my right than to the one who honors me by being on my left; I follow my leaning, I obey my inclination, I abandon myself to the love which she inspires in me; that, ladies, is attraction."[85] Although the exact Newtonian nature of this explanation seems to have been lost, there is no denying that Pilâtre de Rozier spoke directly to the men and women in an engaging and dramatic manner.[86] In fact, in some ways this example typifies how the spectacular side of popular science could get in the way of the theories and ideas behind the demonstrations. Since we do not know if he followed this analogy with a more precise account of gravity, it is unclear exactly what Pilâtre de Rozier's audience would have taken away from this explanation.

This leads us to ask the question of who exactly was in the room? Who constituted this scientific public sphere? By combining enlightened ideas with popular theatrics, and by drawing on salons, academies, and public courses as models for his *musée*, Pilâtre de Rozier enabled his new institution to draw an extraordinarily diverse group of people. The Musée de Monsieur enjoyed the membership of a broad range of savants and amateurs, and both upper- and middle-class individuals could afford to join. To begin with, Pilâtre de Rozier attracted many social and scientific elites to his club, some of whom counted among the most famous and important people of the late eighteenth century. A number of nobles, including the official sponsors, the count and countess of Provence, Louis XVI and Marie-Antoinette, and the duke of Chartres, were official members and patrons of Pilâtre de Rozier's *musée*. Members from the scientific elite included academicians like Jean-Joseph Süe, Félix Vicq-d'Azyr, Fourcroy, and Condorcet, all of whom occasionally taught courses as well.[87] Pilâtre de Rozier ensured that a large number of savants would join by allowing any member of a royal academy to enroll without having to go through the formal admission process. Thus, anyone belonging to the Académie Française, the Académie des Inscriptions & Belles-Lettres, the Académie Royale des Sciences, or the Académie Royale de Médecine had easy access to membership in the Musée de Monsieur. At the same time, Pilâtre de Rozier took advantage of the schism that had divided Court de Gébelin's *musée* by allowing any member of the Musée de Paris to transfer membership automatically to the Musée de Monsieur.[88] There were many people who were upset with the interior squabbles of the Musée de Paris and quickly jumped ship for the relative harmony of the Musée de Monsieur.[89]

However, the Musée de Monsieur should not be considered an elite scientific institution. On the contrary, it made room for savants from every part of the scientific spectrum. Mixed among the nobility and elite philosophers, for example, were several scientific popularizers who joined Pilâtre de Rozier's *musée* seemingly without any objections concerning their social standing or academic credentials. These popularizers included such middling savants as François Bienvenu, an instrument maker and public lecturer who specialized in electricity and magnets; the abbé Guyot, author of the *Nouvelles récréations physiques et mathématiques*, a book of physics experiments and games designed to educate and entertain amateurs; Rouland, a popular demonstrator of experimental physics in the Paris colleges; the abbé Laurent-Antoine Miollan, a public lecturer and aeronaut, later run out of Paris by an angry mob after a failed attempt to launch a balloon; and Charles Deslon, the most prolific and vocal of Mesmer's disciples.[90] All that was required of prospective members, in addition to the annual fee, was the sponsorship of three current members.[91] Under these conditions, the popularizers gained equal admittance with the rest of the savants.

In addition to opening its doors to science popularizers, the Musée de Monsieur also encouraged other non-scientists from the middling level of society to join its growing membership. In fact, one prospectus claimed that "people of any age and of any social status" could spend their leisure time agreeably and usefully by joining the Musée de Monsieur. The official regulations concurred, stating that "all honest persons" could join the *musée*, where they might enjoy a good conversation, attend a lecture, or perform their own experiments.[92] Most advertisements for membership emphasized this openness and claimed that individuals of any age and any social status could join. In reality, all social classes did not and could not join the Musée de Monsieur. If nothing else, the annual fee, set at seventy-two livres for men and thirty-six livres for women, would have excluded most lower-class urban workers or sans-culottes from even considering the possibility. The middling sorts, however, were well represented. Lawyers, secretaries, army officers, minor government officials, architects, teachers, a banker, a law clerk, a publisher, a grocer, and several shop owners all appear in the membership lists.[93] Several of them even sat on the governing council.

Significantly, Pilâtre de Rozier singled out women as important members of the Musée de Monsieur.[94] The few available figures suggest a female membership ranging from ten to twenty percent.[95] Since women had not been able to join La Blancherie's or Court de Gébelin's *musées* as equal members, they were understandably suspicious of Pilâtre de Rozier's offer of equivalent participation. They thought that their reduced membership dues, half that required of men, would limit their access and force them into a position where they were only supporting masculine efforts without really being able to share actively in the advantages of the Musée de Monsieur. The reduced subscription fee for women was not uncommon for public lecture courses, on which Pilâtre de Rozier based the structure of his *musée*, and he was really just adapting a trend common among popularizers. To assuage the fears of women, Pilâtre de Rozier sent an open letter to the editors of the *Journal de Paris* claiming that, on the contrary, women would share identical rights and privileges with men. The ladies would be able "to enjoy the same advantages . . . to which they will have the right" along with the other amateurs.[96] In fact, Pilâtre de Rozier wanted to make the sciences agreeable to women, in much the same way that popularizers wrote books with a female audience in mind. They were seen as the perfect audience: women, he wrote, "prove to us every day the progress of our century."[97]

The degree of participation allowed women in the Musée de Monsieur sets it apart from other associations and, indeed, other *musées*. But when seen in light of public lecture courses and women's participation in those venues, their equal acceptance in the Musée de Monsieur seems less remarkable. The same can be said for having a mixed audience of men and women, another typical scenario in public lectures. The inclusion of both women and men in the expanding audience for the consumption of science in the public sphere provided an important impetus for the popularization of science and the creation of Enlightenment culture.

In sum, the *musées* offered multiple opportunities for men and women from the middling and upper classes. At the same time the *musées* contributed to the creation of scientific culture by establishing a concrete location where Parisians could go to learn about the most recent discoveries and learn how to apply science to their lives. This mixture of elite and middling savants, upper and middle classes created an arena where individuals could meet on relatively equal footing to learn about and discuss current scientific trends and ideas. Individuals joined the *musée* to gain access to the kind of scientific cultural capital that had become particularly popular during the late eighteenth century. Science was not only an exclusive and elite activity practiced by nobles and philosophes but had become an important part of urban culture.[98]

The *musées* embodied and made more widely accessible many of the ideals formulated and institutionalized by the *Encyclopédie*.[99] By taking all the different public lecture courses available on the streets of Paris and placing them in one location, the founders of the *musées* provided their members easy and complete access to the Enlightenment project in one convenient location and for one annual fee. The dues paid to *musées* were less than what one would have paid to purchase all the books and equipment found there and certainly saved members money compared to the sum total of the individual public lectures.

In these clubs, the Enlightenment itself was commodified and packaged for consumption by the general Parisian public in the same way public lecture courses had been doing in Paris since the 1730s. Prosperity in the flooded market of the *musées* depended on attracting members and providing a good product, in terms of both the physical space and the perceived usefulness of the available resources. Pilâtre de Rozier enjoyed particular success by blending enlightened ideas with his commercial and advertising skills. To do this, he drew on his own experiences as a public lecturer and his talent for combining utility and entertainment.[100] The encyclopedic breadth of the *musées* testifies to their kinship to public lecture courses. They formalized the system of popularization into a specific public space, thus giving their members access to the pedagogical revolution that had been sweeping Paris since Jean-Antoine Nollet first started teaching classes during the 1730s. Never could so much Enlightenment, in its various forms, be attained in one location and for such a low price. This appropriation of a useful and entertaining pedagogy, combined with a socially and economically diverse audience willing and able to participate in the consumption of Enlightenment science made the arena of the *musées* distinct. The *musées*, like public lecture courses, salons, or freemason lodges, institutionalized the Enlightenment. Their members used the *musées* to spread scientific ideas and to create a popular scientific culture.

Notes

1 "Mémoires de J.-P. Lenoir," Bibliothèque municipale d'Orléans, ms. 1423, f. 235. I would like to thank Alan Williams and Dena Goodman for generously making this manuscript available to me. The creation of new clubs and societies occurred throughout Europe. See James E. McClellan, *Science Reorganized: Scientific Societies in the Eighteenth Century* (New York: Columbia University Press, 1985), 151; and Richard van Dülmen, *The Society of the Enlightenment: The Rise of the Middle Class and Enlightenment Culture in Germany*, trans. Anthony Williams (New York: St. Martin's Press, 1992).

2 Luc-Vincent Thiéry, *Guide des amateurs et des étrangers voyageurs à Paris*, 2 vols. (Paris: Hardouin & Gattey, 1786–1787), I:232–5.

3 Mederic-Louis-Elie Moreau de Saint-Méry, *Discours sur l'utilité du musée établi à Paris; Prononcé dans sa séance publique du 1 décembre 1784* (Parma: Bodoni, 1805), 8–9.

4 Fifteen livres would be about one week's wages for a well-off artisan. On wages see Jeffrey Kaplow, *The Names of Kings: The Parisian Laboring Poor in the Eighteenth Century* (New York: Basic Books, 1972), 53–4; George Rudé, "Prices, Wages and Popular Movements in Paris During the French Revolution," *Economic History Review* 6 (1954), 248; and Jean Sgard, "L'Echelle des revenus," *Dix-huitième siècle* 14 (1982): 425–33.

5 Daniel Roche, *Le Siècle des lumières en province: académies et académiciens provinciaux, 1680–1789*, 2 vols. (Paris: Mouton, 1978), I:66. Also, Daniel Mornet, *Les origines intellectuelles de la Révolution Française, 1715–1787* (Paris: Armand Colin, 1967 [1933]), 284–5.

6 Goodman, *The Republic of Letters: A Cultural History of the French Enlightenment* (Ithaca: Cornell University Press, 1994). On the *musées* also see Hervé Guénot, "Musées et lycées parisiens (1780–1830)," *Dix-huitième siècle* 18 (1986): 249–67.

7 For a critique of Goodman's argument, relative to freemason lodges, see Janet M. Burke and Margaret C. Jacob, "French Freemasonry, Women, and Feminist Scholarship," *Journal of Modern History* 68 (1996), 546. They claim that women's freemason lodges were fashioned

directly upon the foundations created by salons and, therefore, that they represented a significant location for female participation, and governance, in Enlightenment culture.

8 See Guénot, "La Correspondance générale pour les Sciences et les Arts de Pahin de La Blancherie (1779–1788)," *Cahiers Haut-Marnais* 162 (1985): 49–61 as well as his "Musées et lycées parisiens", and his "Une nouvelle sociabilité savante: Le Lycée des Arts," in *La Carmagnole des muses: L'Homme de lettres et l'artiste dans la Révolution* (Paris: Armand Colin, 1988): 67–78.

9 See Charles Cabanes, "Histoire du premier musée autorisé par le gouvernement," *La Nature* (1937): 577–83; and William A. Smeaton, "The Early Years of the Lycée and the Lycée des Arts," *Annals of Science* 11 (1955): 257–67, 309–19.

10 See Louis Amiable, "Les origines maçonniques du Musée de Paris et du Lycée," *La Révolution Française* 31 (1896): 484–500; and Charles Dejob, "De l'éstablissement connu sous le nom de Lycée et d'Athénée et de quelques établissements analogues," *Revue internationale de l'enseignement* 18 (1889): 4–38.

11 On the creation of various institutions of Enlightenment sociability see Jürgen Habermas, *The Structural Transformation of the Bourgeois Public Sphere*, trans. Thomas Burger (Cambridge, MA: MIT Press, 1989), esp. 14–43; Reinhart Koselleck, *Critique and Crisis: Enlightenment and the Pathogenesis of Modern Society* (Cambridge, MA: MIT Press, 1988); and Dena Goodman, "Public Sphere and Private Life: Toward a Synthesis of Current Historiographical Approaches to the Old Regime," *History and Theory* 31 (1992): 1–20.

12 On salons see Goodman, *The Republic of Letters*; cf. Alan C. Kors, *D'Holbach's Coterie: An Enlightenment in Paris* (Princeton: Princeton University Press, 1976). For cafés see Thomas Brennan, *Public Drinking and Popular Culture in Eighteenth-Century Paris* (Princeton: Princeton University Press, 1988). On reading societies see Paul Benhamou, "La lecture publique des journaux," *Dix-huitième siècle* 24 (1992): 283–95; and P. Benhamou, "The Reading Trade in Lyons: Cellier's *cabinet de lecture*," *Studies on Voltaire and the Eighteenth Century* 308 (1993): 305–21. On French freemasonry see Janet M. Burke, "Freemasonry, Friendship and Noblewomen: The Role of the Secret Society in Bringing Enlightenment Thought to Pre-Revolutionary Women Elites," *History of European Ideas* 10 (1989): 283–93; and Burke and Jacob, "French Freemasonry." For a Europe-wide look see Jacob, *Living the Enlightenment: Freemasonry and Politics in Eighteenth-Century Europe* (Oxford: Oxford University Press, 1991). On provincial academies see Roche, *Le Siècle des lumières en province*; and for Paris see Roger Hahn, *The Anatomy of a Scientific Institution: The Paris Academy of Sciences, 1666–1803* (Berkeley: University of California Press, 1971).

13 Louis Sébastien Mercier, *Les Entretiens du Palais Royal* (Utrecht: Buisson, 1786): 109–10.

14 Darnton, "The High Enlightenment and the Low-Life of Literature," in *The Literary Underground of the Old Regime* (Cambridge, MA: Harvard University Press, 1982): 1–40.

15 Of course, the Enlightenment did not address itself to every level of society; peasants and the urban poor, for example, would have been almost automatically excluded from this scenario. See Harvey Chisick, *The Limits of Reform in the Enlightenment* (Princeton: Princeton University Press, 1981).

16 On the subject of cultural capital see Pierre Bourdieu, *Distinction: A Social Critique of the Judgement of Taste*, trans. Richard Nice (Cambridge, MA: Harvard University Press, 1984).

17 On knowledge as a commodity in England, see John Money, "Teaching in the Market-Place, or 'Caesar adsum jam forte: Pompey aderat': The Retailing of Knowledge in Provincial England during the Eighteenth Century," in *Consumption and the World of Goods*, eds. John Brewer and Roy Porter (London: Routledge, 1993): 335–77.

18 On Pahin de la Blancherie see Felix Rabbe, "Pahin de la Blancherie et le salon de la correspondance," *Bulletin de la Société Historique du VIe Arrondissement de Paris* 2 (1899): 30–52; and Hervé Guénot, "Pahin de la Blancherie," in *Dictionnaire des journalistes*, ed. Anne-Marie Chouillet, supp. 2 (Grenoble: Presses universitaires de Grenoble, 1983): 168–76.

19 Bachaumont, *Mémoires secrets pour servir à l'histoire de la République des Lettres en France depuis 1762 jusqu'à nos jours*, 36 vols. (London: John Adamson, 1784–1789), 26 November 1779.

20 Guénot, "La Correspondance générale pour les Sciences et les Arts."
21 *Nouvelles de la République des Lettres et des Arts*, 6 July 1779, 163.
22 *Nouvelles*, 1 June 1779, 127. See also the *Prospectus*, in *Nouvelles*, 3–9.
23 A description of one of the early meetings of the Salon de la Correspondance can be found in François-Valentin Mulot, "Journal intime de l'abbé Mulot," ed. Maurice Tourneux, *Mémoires de la Société de l'histoire de Paris et de l'Ile-de-France* 29 (1902), 47–50.
24 The *Nouvelles*, patterned after Pierre Bayle's journal, ran from 1779 to 1786. See Guénot, "Les Lecteurs des *Nouvelles de la République des Lettres et des Arts* (1782–1786)," in *La Diffusion et la lecteur des journaux de langue française sous l'ancien régime*, ed. Hans Bots (Amsterdam & Maarssen: APA-Holland University Press, 1988): 73–88; and Jean Sgard, *Dictionnaire des journaux*, 2 vols. (Paris: Universitas, 1991), II:947–8.
25 Bachaumont, *Mémoires secrets*, 22 April 1782.
26 *Nouvelles*, 1 June 1779, 127.
27 The correspondance between John Adams and Thomas Jefferson is collected in *The Adams-Jefferson Letters: The Complete Correspondance Between Thomas Jefferson and Abigail and John Adams*, ed. Lester J. Cappon, 2 vols. (Chapel Hill: University of North Carolina Press, 1959), I:196, 212, 222, and 226.
28 Guénot, "La Correspondance générale," 59.
29 *Nouvelles*, 20 April 1779, 74–76; ibid., 4 May 1779, 89 (Guyot's pump); ibid., 28 December 1779, 63 (Dellabarre's microscope); ibid., 25 January 1780, 99 (Girardin's electrical machine); ibid., 26 June 1782, 190 (weather predictor).
30 "Extrait des Registres de l'Académie Royale des Sciences de Paris, du 20 May 1778," in Pahin de la Blancherie, *Correspondance générale sur les sciences et les arts* (Paris: Au bureau de la Correspondance, 1779), 27–8.
31 *Mémoires secrets*, 21 November 1786.
32 *Mémoires secrets*, 1 July 1781. On the lottery and La Blancherie's money problems see Goodman, *The Republic of Letters*, 250.
33 *Mémoires secrets*, 9 December 1782; ibid., 18 December 1785; ibid., 10 December 1781; ibid., 2 March 1785; and ibid., 4 February 1786.
34 *Correspondance littéraire, philosophique et critique*, ed. Maurice Tourneux, 16 vols. (Paris: Garnier frères, 1877–1882), May 1778, XII:103.
35 *Mémoires secrets*, 28 October 1782.
36 The claim for ninety-four noble members can be found in *Mémoires secrets*, 28 October 1782. For a list of forty-one of the noble members see *Mercure de France* 30 June 1781, 236–7. The count of Provence was the club's chief patron although he would later become the patron of other organizations as well: see *Mémoires secrets*, 22 April 1782.
37 For membership numbers see *Mémoires secrets*, 28 October 1782. A different set of numbers can be found in the *Nouvelles*, 15 August 1782, 242–5. There, ninety-one protectors are listed along with fifty-five associates. For La Blancherie's comments on the subscriber class see *Mercure de France*, 20 November 1779, 135. On the mixture of social groups see Richard Derb . . . to the Editor of the *Courier de l'Europe*, 8 May 1779, quoted in Goodman, *The Republic of Letters*, 248.
38 By 1782 only a handful of women had officially joined as protectors. See *Nouvelles*, 15 August 1782, 242–5.
39 *Mercure de France*, 20 November 1779, 132; and ibid., 30 June 1781, 235.
40 Goodman, *The Republic of Letters*, 234, 246 fn. 38; *Mémoires secrets*, 24 November 1783.
41 Little has been written on the Musée de Paris. A notable exception is Sophia A. Rosenfeld who discusses it as an organization that provided experimental space for the study of gestural communication: see *A Revolution in Language: The Problem of Signs in Late Eighteenth-Century France* (Stanford: Stanford University Press, 2001), 110–12, 116–21. On Court de Gébelin see Paul Schmidt, *Court de Gébelin à Paris (1763–1784)* (Saint-Blaise: Foyer Solidariste, 1908); Justin Cabrière, *Court de Gébelin, défenseur des églises réformées de France, (1763–1784)* (Paris: Cahors, 1899); and Jean-Paul Rabaut-Saint-Etienne, "Lettre sur la vie et les écrits de

M. Court de Gébelin, adressée au Musée de Paris," in *Oeuvres de Rabaut-Saint-Etienne*, 2 vols. (Paris: Laisne, 1826): II:355–90.

42 *Mémoires de Musée de Paris* (Paris: Moutard, 1784–1785), I:ii, ix.

43 *L'Année littéraire*, 1786, V:228.

44 *Mémoires secrets*, 7 December 1782.

45 Some modern scholars have latched on to this criticism as well. Darnton, for example, writes that "the musées of Court de Gébelin and P.C. de La Blancherie seem even to have served as counteracadamies and antisalons for the multitude of philosophes who could not get a hearing elsewhere." See "The High Enlightenment and the Low-Life of Literature," *The Literary Underground of the Old Regime* (Cambridge, MA: Harvard University Press, 1982), 24. Darnton, however, draws much of his information from the *Mémoires secrets*, a journal that was decidedly against the *musées* in general. Cf. Wallace Kirsop, "Cultural Networks in Pre-Revolutionary France: Some Reflexions on the Case of Antoine Court de Gébelin," *Australian Journal of French Studies* 18 (1981): 231–47, esp. 234–40.

46 *Mémoires secrets*, 10 March 1782.

47 La Vilemarias to Seignette, 12 February 1783, cited in Darnton, *Mesmerism and the End of the Enlightenment in France* (Cambridge, MA: Harvard University Press, 1968), 178.

48 On the schism in the Musée de Paris see *Mémoires secrets*, 27 July 1783, 7 August 1783, 9 August 1783, 1 January 1784, and 28 January 1784. The schism resulted in two separate musées both claiming the title of Musée de Paris.

49 *Mémoires secrets*, 1 January 1784; and ibid., 5 June 1784.

50 Brissot claimed to be quoting the philosophe Claude Helvétius. Brissot de Warville, *Mémoires de Brissot*, 4 vols., ed. M.F. de Montrol (Paris: Ladvocat, 1830): I:76.

51 *Mémoires secrets*, 1 March 1783. This is one of the few overtly political acts performed by any of the *musées*. I suspect that the Musée de Paris took this step largely because of its Masonic connections, Court de Gébelin also being a member of the Loge des Neuf-Soeurs along with Franklin.

52 Luc-Vincent Thiéry, *Almanach du voyageur à Paris*, 5 vols. (Paris: Hardouin, 1783–1787), II:448–9.

53 See the poem praising the female honorary associates of the Musée de Paris in the *Mercure de France*, 14 February 1784, 53.

54 Court de Gébelin in a letter to a friend, quoted in Maurice Pellisson, *Les Hommes de lettres au XVIIIe siècle* (Paris: Armand Colin, 1911), 212–13.

55 *Mémoires du Musée de Paris*, II:xviii.

56 M. Ducarla, for example, offered a course on the nature of fire: *Journal de Paris*, 23 December 1784, 1514. A written version of this class was published in the *Mémoires du Musée de Paris*, vol. I (Paris: Moutard, 1784).

57 Duplessy, *Répertoire des lectures faites au Musée des Dames* (Paris: Cailleau, 1788), 3.

58 Women could, for example, join some freemason lodges and often did so for reasons similar to those of women who joined the *musées*. Burke and Jacob note that lodges "broadened and deepened the experience of enlightenment culture" for women. See Burke and Jacob, "French Freemasonry," 546.

59 Duplessy, *Répertoire*, 11–12, 20–21.

60 Arbeltier, *Lycée des femmes. Plan de cet établissement* (Paris: L'Imprimerie de Roland, n.d.), 8–12.

61 The Lycée des Arts, frequently confused with the Musée de Monsieur which, during the French Revolution, also used the name Lycée or Lycée Républicain, was founded in 1792. On this club see Guénot, "Une nouvelle sociabilité savante"; *Lycée des arts: Prospectus* (Paris: Au local du Lycée des Arts, Jardin Egalité, An III); Lucien Scheler, *Lavoisier et la Révolution française*, vol. I, *Le Lycée des arts* (Paris: Hermann, 1956); and Smeaton, "The Early Years of the Lycée and the Lycée des Arts."

62 Several short histories of this *musée* have been written: see Dejob, "De l'établissement connu sous le nom de Lycée," 4–38; Cabanes, "Histoire du premier musée"; and Smeaton, "The

Early Years of the Lycée."

63 *Mémoires secrets*, 17 February 1782, 79–80. In this case the target was Pahin de la Blancherie's General Assembly.

64 For the sake of convenience, I will refer to this club as the Musée de Monsieur regardless of these numerous name changes.

65 *Musée, autorisé par le gouvernement, sous la protection de Monsieur et de Madame* (Paris: N.p., n.d.); and *Affiches de Paris*, 4 December 1781, 2790–1. The oldest brother of the King was usually called "Monsieur." On the organization of the musée see *Statuts et règlemens du premier musée autorisé par le gouvernement* (N.p., n.d.); and *Conseil du premier musée* (N.p., n.d.). For financial information see "Mémoire d'un abonné à Messieurs les souscripteurs du Lycée," and "Etat des dépenses du Lycée," in AN, T427 and AN, MC – XLII, 640.

66 Smeaton, *Fourcroy: Chemist and Revolutionary, 1755–1809* (Cambridge: W. Heffer & Sons, Ltd., 1962), 16. For the history of the Musée de Monsieur in the nineteenth century see Barbara Haines, "The Athénée de Paris and the Bourbon Restoration," *History and Technology* 5 (1988): 249–71; and Martin Staum, "Physiognomy and Phrenology at the Paris Athénée," *Journal of the History of Ideas* 56 (1995): 443–62.

67 For biographical information on Pilâtre de Rozier see Antoine Tournon de la Chapelle, *La Vie et les mémoires de Pilâtre de Rozier, Ecrits par lui-même* (Paris: Belin, 1786); Léon Babinet, "Notice sur Pilâtre de Rozier," *Mémoires de l'Académie nationale de Metz* 46 (1865): 161–228; Paul Dorveaux, "Pilâtre de Rozier," *Bulletin de la Société d'Histoire de la Pharmacie* (1920): 249–58; Clément Duval, "Pilâtre de Rozier (1754–1785): Chemist and First Aeronaut," *Chymia* 12 (1967): 99–117; and William A. Smeaton, "Jean-François Pilâtre de Rozier, The First Aeronaut," *Annals of Science* 11 (1955): 349–55.

68 On the Société Libre d'Emulation see Goodman, *Republic of Letters*, 260–3.

69 *Musée, autorisé par le gouvernement* (n.p., n.d.), 3. He also held the position of "Secretary of Madame's Cabinet" although it is unclear what duties he performed in that capacity.

70 Concerning a state commission, this one dealing with the problem of noxious airs in Paris, see "Extrait du Rapport des Expériences faites en présence des commissaires nommés par le Gouvernement pour constater la Découverte de M. Pilâtre de Rozier sur le Méphitisme," AN, O1–1293, f. 40; Paul Dorveaux, "Pilâtre de Rozier et l'Académie des Sciences," *Les Cahiers lorrains* 8 (1929): 162–6, 182–5; and Alain Corbin, *The Foul and the Fragrant: Odor and the French Social Imagination* (Cambridge, MA: Harvard University Press, 1986), 94. For some of his publications see for example *Journal de Physique*, 20 (1782): 351–61 and ibid., 16 (1780): 381–8.

71 See for example *Affiches de Paris*, 6 July 1782, 1566; *Journal de Paris*, 4 March 1782, 249–50; and ibid., 2 January 1782, 5–6.

72 For Pilâtre de Rozier's voyage in the hot-air balloon, see his account in *Première expérience de la Montgolfiére, construite par ordre du roi* (Paris: L'Imprimerie de Monsieur, 1784). On his death in an ill-fated attempt to cross the English Channel from France to England see Charles Cabanes, "La mort d'Icare: Pilâtre de Rozier," *La Nature* (1936): 529–33; and Smeaton, "The First and Last Balloon Ascents of Pilâtre de Rozier," *Archives internationales d'histoire des sciences* 44 (1958): 263–9.

73 The Musée de Monsieur initially had rooms on the Rue Sainte Avoie, near the Rue de Plâtre, before moving to the more fashionable Rue Saint-Honoré near the Palais Royal.

74 For a description of the rooms at the first location see *Musée, autorisé par le gouvernement* (N.p., n.d.), 2–3. For the new rooms near the Palais Royal see the depiction in "Lycée républicain, Etat de mobilier," in AN, F17–1331B, d. 9, f. 189; and John G. Lemaistre, *A Rough Sketch of Modern Paris*, 2nd edn (London: Printed for J. Johnson, in St.-Paul's Churchyard, 1803), 255–6.

75 *Mémoires secrets*, 9 September 1783. See also *Affiches de province*, 25 December 1783, 218–19.

76 *Statuts et règlemens*, 1–2, 4.

77 *Programme du lycée, pour l'année 1788* (N.p., n.d.), 1, 14–15.

78 The after-death inventory of Pilâtre de Rozier's possessions included well over 300 scienti-

fic instruments: AN, MC XLII, 624; for an inventory from 1792 see AN, F17–1331B, d. 9, f. 187 and f. 189.

79 *Journal de Paris*, 27 November 1781, 1332.

80 *Musée autorisé par le gouvernement, sous la protection de Monsieur et de Madame* (Paris: L'Imprimerie de Monsieur, 1784), 4.

81 *Musée autorisé par le gouvernement* (n.p., n.d.), 3; and *Journal de Paris*, 7 February 1786, 153.

82 Marie Jean Antoine Nicolas de Caritat, Marquis de Condorcet, "Discours sur les sciences mathématiques, prononcé au Lycée," in *Oeuvres*, eds. A. Condorcet O'Conner and M.F. Arago, 12 vols. (Paris: Firmin Didot frères, 1847), I:472, 473.

83 Condorcet, "Discours sur l'astronomie et le calcul des probabilités lu au Lycée en 1787," in *Oeuvres*, I:502–3.

84 Moreau de Saint-Méry, *Discours sur l'utilité du musée établi à Paris*, 9, 30.

85 Tournon de la Chapelle, *La vie et les mémoires de Pilâtre de Rozier*, 11–12. This scientization of love and the equating of gravitational attraction with sexual attraction foreshadows Charles Fourier's utopian vision of passionate attractions expounded in the first part of the nineteenth century. See Charles Fourier, *The Utopian Vision of Charles Fourier*, trans. and ed. Jonathan Beecher and Richard Bienvenu (Columbia, MO: University of Missouri Press, 1983).

86 As with certain books popularizing science, like Bernard de Fontenelle's *Conversations on the Plurality of Worlds* (Berkeley: University of California Press, 1990) or Francesco Algarotti's *Newtonianism for Ladies*, Pilâtre de Rozier addressed himself to the women in the audience although the lectures, like the books, were intended for a mixed audience.

87 Many well-known nineteenth-century savants, including Georges Cuvier, Benjamin Constant, and August Comte, were members of the Musée de Monsieur. See Louis Amiable, "Les origines maçonniques du Musée de Paris et du Lycée," *La Révolution française* 31 (1896), 500.

88 *Statuts et règlemens*, 22.

89 Pierre-Jean-Baptiste Nougaret, *Tableau mouvant de Paris, ou, Variétés amusantes*, 3 vols. (Paris: Duchesne, 1787), I:230–1; *Mémoires secrets*, 1 January 1784, XXV:13.

90 *Liste de toutes les personnes qui composent le premier musée autorisé par le Gouvernement, sous la protection de Monsieur et de Madame* (Paris: Imprimé par Ordre du Conseil du Musée, 1785), 3–46: AN, MC XLII, 628. A copy of this can also be found at the Bibliothèque historique de la ville de Paris (book code, 32,002).

91 *Statuts et règlemens*, 21. The annual fee was 72 livres although the price later went up to 96 livres: *Programme du Lycée, pour l'année 1788*, 13.

92 Of course, the term "honest" implied a distinct stature and was associated with a certain level of education, civility, and politeness, but this phrase only occurs in the statutes; most of the advertisements imply a more open membership policy. *Journal de Paris*, 7 February 1786, 153; *Statuts et règlemens*, 21; *Musée de Monsieur, et de Monseigneur Comte d'Artois* (Paris: L'Imprimerie de Monsieur, 1785), 3.

93 *Liste de toutes les personnes*, 3–46. Also see AN, F17–1331B, d. 9, f. 190; and AN, MC XLII, 640.

94 *Journal encyclopédique*, January 1788, 1112.

95 A subscription list for 1785 lists 727 total members of whom 77 were women. Another list, compiled during the French Revolution, includes 196 total members of whom 38 were women. See AN, F17–1331B, d. 9, f. 190 and f. 191.

96 *Journal de Paris*, 12 December 1781, 1391–2. The only aspect women could not participate in were the evening meetings, which seem to have been organizational in nature. So while women were allowed free access to most parts of the *musée* and were allowed to become founding members (if they donated enough money), they could not hold office.

97 *Musée de Monsieur* (1785), 32.

98 This growth in the market of ideas mirrors activities that occurred in other cultural arenas at the end of the eighteenth century. See Daniel Roche, *The Culture of Clothing: Dress and*

Fashion in the "Ancien Régime," trans. Jean Birrell (Cambridge: Cambridge University Press, 1994), 496; Roche, *A History of Everyday Things: The Birth of Consumption in France, 1600–1800*, trans. Brian Pearce (Cambridge: Cambridge University Press, 2000); and Cissie Fairchilds, "The Production and Marketing of Populuxe Goods in Eighteenth-Century Paris," in *Consumption and the World of Goods*, 228–48.

99 Goodman, *The Republic of Letters*, 242, 266.

100 Moreau de Saint-Méry, *Discours sur l'utilité du musée*, 1–2.

5

Divining rods and public opinion

At 10:00 p.m. on 5 July 1692 thieves broke into the Lyon wine shop owned by Antoine Boubon Savetier and his wife, bludgeoned them to death with a bill-hook, and escaped with approximately five hundred livres.[1] When the local authorities made no progress on the case a wine dealer from Dauphiné stepped forward and recommended the services of Jacques Aymar, a peasant well known for having solved an equally difficult murder case. With little choice the authorities called on Aymar's help. He arrived in Lyon, inspected the site of the murder, and immediately started off on the trail of the culprits. Aymar first led police out of Lyon and down the Rhône River where he tracked the killers to the home of a gardener. Once there Aymar confidently announced that three people had committed the crime. He indicated the table where the murderers had sat, and pointed out the wine bottle they had used, information corroborated by the gardener's two young children.[2] Aymar then led the police to the nearby town of Beaucaire where he followed the trail to the local jail and directly to one of the inmates who just one hour earlier had been arrested for petty larceny. This man, Joseph Arnoul, a nineteen-year-old from Toulon and easily identified because of a hunchback, at first denied the accusations and said he had never been to Lyon. Nonetheless, the guards arrested him and took him back to the scene of the crime where witnesses, including the gardener's children, identified him and claimed they had seen him loitering near the wine merchant's shop. At this point Arnoul confessed and named his two accomplices, one called Thomas and the other André Pese, both notorious criminals based in Toulon.

Aymar went back to work: he returned to Beaucaire and continued following the trail, first to Toulon, where he led police to the very inn where the suspects recently had dined, and then to the sea, where apparently they had boarded a ship. Undeterred, Aymar obtained his own vessel and continued to track them along the coast of France until it became clear they were making for Genoa. Since he and his police escort lacked the authority to make an arrest in a foreign city they decided to return home. For his part, Joseph Arnoul claimed that his accomplices had committed the actual murder. Nonetheless, he was tried, found guilty, and executed by being broken on the wheel on 30 August 1692.[3]

Jacques Aymar's spectacular feat of detection made him an instant celebrity; and it immediately sparked a huge controversy. The reason for the dispute centered around the fact that Aymar had tracked down the killers with a divining rod, a forked stick usually used in the dowsing trade to find underground springs and ores.[4] By itself, the discussion sparked by Aymar's solution to the double murder provides a unique perspective on the creation of public opinion during the early stages of the age of Enlightenment. The controversy surrounding Aymar's methods ultimately involved doctors, theologians, natural philosophers, representatives of the state, and even astrologers. The resulting mixture of ideas and notions illustrates the range of cultural and intellectual tools available at the end of the seventeenth century and at the beginning of the Enlightenment.

The contention over dowsing, however, offers a more revealing portrait of the period when juxtaposed next to a similar debate held in the 1770s and 1780s over the capabilities of Barthelemy Bléton, another Dauphinoise peasant whose talent with a divining rod captured the imagination of the French. In this case the quarrel over Bléton, who had discovered springs for Queen Marie-Antoinette at Versailles and occasionally found coalfields for the state, took place among savants and members of the general public. This battle developed over the proper location of scientific authority and the utility of natural philosophy in everyday life. Just as the Aymar case corresponds neatly with the origins of the Enlightenment, the Bléton case fits well with the end, or at least the culmination, of the Enlightenment during the 1780s.[5]

This chapter will follow the shifting terms of the debate over rabdomancy (the official, and somewhat scientific, name for the art of dowsing). During that time period the form, the forum, and the players in the controversy all shifted to accommodate the growing public interest in scientific matters and the appropriation of natural philosophy into popular culture. This shift illustrates the evolution of a scientifically-aware public sphere during the eighteenth century characterized both by the plethora of social and cultural arenas in which discussions of dowsing took place and by the incredible growth of the audience who contributed to those discussions. Not only did the number of people able and willing to take part actively in this argument grow dramatically during the eighteenth century, but those people also represented a much broader range of social, educational, and economic backgrounds. The changing nature of the participants grew partially from changing attitudes towards natural philosophy, the growth in public participation in popular science, and the ongoing

Enlightenment stress on the interconnections between science and utility. Thus, while balloons, discussed in the next chapter, achieved recognition as scientifically important in spite of the difficulty in determining what use they might have, dowsing appeared eminently practical while remaining scientifically suspect.

As the general public appropriated the tools and language of enlightened discourse they also came to take on and alter its purpose. The rise in importance of public science transformed the role of the public in scientific debates. Throughout the eighteenth century more and more people placed more and more emphasis on reason and utility; but it was frequently a singularly personal version of these notions, one that could ignore the wishes and desires of the intellectual and cultural elite. Many of the people who involved themselves in public controversies, such as the debate over dowsing, were not those whom the century's leading savants would have chosen. The interested but relatively unlearned amateur public placed considerably more emphasis on utility and results than they did on theoretical unity. In the dispute over rabdomancy we find that on one side the voice of reason and rationality came forth from various savants. On the other side we have a population willing to accept and follow a course of reasoning based on "what worked" rather than on what had philosophical substance. People backed dowsing, oftentimes against the desires of members of the state and especially the state's spokespeople and legitimizers of good science, the academicians, because dowsers fulfilled their claims and provided a very useful service. As P.J.C. Mauduyt de la Varenne noted, if savants expressed two opposing opinions then experience alone should be the judge; unfortunately, from the point of view of the elite savants, experience showed that, more often than not, dowsing worked.[6] In this instance, the public, it turns out, had the courage to make use of its own reason, much to the chagrin of those who composed the anti-dowsing brigade. Savants, lawyers, and philosophes did not hold a monopoly on public opinion; sometimes the public came forward with their own voice thanks to the appropriation of science in urban culture during the eighteenth century. The growing participation of people in public opinion debates was not limited to science but found expression in a variety of social and intellectual forums from art and science to medicine and politics.[7]

Jacques Aymar, dowsing detective

The art of dowsing, or the science of rabdomancy depending on one's point of view, has existed since ancient times, in a number of rather vague forms. Proponents of dowsing often point to Jacob's rod, as well as that of Moses, as possible early examples of divining rods. The first recorded appearance of this practice in early modern Europe came during the fifteenth and sixteenth centuries when German miners apparently practiced dowsing in order to locate appropriate places to dig. The practice spread from the Holy Roman Empire to England and the rest of the continent. Dowsing underwent its first extended systematic analysis at the hands of Martine de Bertereau, baronne de Beausoleil, in her books, the *Véritable déclaration de la découverte des mines et minières de France* and *La Restitution de Pluton*.[8] The Baroness wisely dedicated the latter book to Armand-Jean du Plessis, Cardinal de Richelieu, in the hopes of gaining patronage for herself and her husband (a mineralogist and an Inspector-General of Mines for France). Rumors circulated, however, concerning the occult nature of her studies and suspicions arose over whether she might be practicing magic, a serious accusation during this time of witch-hunts. This rumor, combined with what turned out to be the bad politics of her husband, brought them the wrong kind of attention from the Cardinal and in 1642 she landed in the prison at the Château de Vincennes where she eventually died.

The sixteenth and seventeenth centuries saw additional studies of dowsing, both for and against. Jacques Le Royer, in his *Traité du bâton universel*, claimed that the divining rod could be used to unearth all sorts of hidden things, a view he shared with savants like Gaspard Schott and Basile Valentin. Pierre Gassendi and Robert Boyle essentially said that practitioners of dowsing could use it to find water and metal deposits, but that they should avoid it for other uses. Georg Agricola, on the other hand, came out against it.[9] While most of these authors focused on the practice of dowsing to find water and ores, it seems that the divining rod also began to find use in a variety of other situations during the seventeenth century. Some scholars, such as Pierre Bayle in his *Dictionnaire historique et critique*, noted that diviners claimed they could track thieves and murderers (Aymar is the example here), discover forgotten land boundaries, find buried treasures, identify the fathers of abandoned children, and determine if a man or a woman had committed adultery or, alternately, if they still retained their virginity.[10] The utility of the divining rod could surely have been quite profound, if only people could agree on how it worked. Aymar's incredible success led to the first full-fledged debate over whether or not rabdomancy actually had a sound scientific basis.

Jacques Aymar was born just after midnight in Saint-Marcellin on 8 September 1662 (the date and exact time held considerable importance for those espousing astrological explanations of dowsing). It is not known how he came to find that he possessed dowsing capabilities, but Dauphiné had a reputation for producing dowsers and it is likely that he initially learned of his abilities through the imitation of others. He first came to realize he could use his talents to solve crimes in 1688 when, while out searching for water, he felt his divining rod turn so sharply that he believed he had found a major spring.

When the workmen dug down, however, they found not water but the body of a local woman, missing for the last four months, buried inside a barrel. The body was still recognizable in spite of the circumstances of its discovery. That she had been murdered was certain since the cord used to strangle her was found with her body.[11] Aymar went to the former home of the murdered woman and directed his rod at each person there; but it only moved towards the widower, who immediately fled, thus establishing both his guilt and Aymar's ability to track criminals. In a later case, this time in Grenoble, Aymar used his divining rod to solve a crime concerning the theft of some clothes. He amazed the local magistrate by tracking down and denouncing the two thieves responsible, both of whom admitted their part in the crime.

With this reputation behind him, the authorities in Lyon called on Aymar's help to solve the murder of the wine merchant and his wife. However, it should be noted that these men were not particularly credulous or gullible; they first put Aymar's talents to the test to satisfy their own concerns over his legitimacy. They buried the murder weapon, along with several similar tools, and asked Aymar to determine not only where they had hidden it, but also which of the several tools was the true weapon. Aymar accomplished this task twice, the second time while blindfolded. With his abilities experimentally demonstrated to the local state representatives, including the head of the police, the local judge, and the Intendant, Pierre de Bérulle, the authorities gave Aymar temporary legal powers and a number of guards to accompany him. He then tracked down the three men responsible for the murders.

The explanatory models put forward to help explain Aymar's success and the practice of rabdomancy in general represents an astounding array of ideas and practices. Some scholars, for example, undertook historical studies (usually combined with medical, scientific, or religious explanations) and collected stories, citations, and anecdotes from sources, both Christian and classical, designed to demonstrate the ancient origins of dowsing. They alternately sought to legitimize or discredit rabdomancy through an elaboration of its antiquity and its association with either Christian or pagan rituals.[12]

Another theory concentrated on the astral influences behind dowsing. Astrologers concentrated their attention on the confluence of stars and planets present at the births of dowsers. As a result, various nativities were cast for Aymar and other dowsers to see if any correlations could be discovered.[13] Astrologers also believed that the positions of the heavens influenced the timing involved when a dowser obtained his or her divining rod. Iron, for example, should be sought using a rod cut under the influence of Mars while diamonds, on the other hand, were under the sway of the Moon.[14] This astrological theory rested on the idea that the arrangement of the planets caused certain individuals, born at the right time, to be more in tune with the ebb and flow of ether on top of and under the earth's crust. In other words, certain people could tap into the chain of being more easily than others based on the arrangement of the heavens at the moment of their birth. In this way, dowsers could literally sense the movement of water, or the placement of ores, or the bad energy of a murderer and, with the aid of an appropriately cut divining rod, trace the movement of that energy flow. A problem arose for the astrologers, however, when they discovered that Aymar was not an Aquarius, as they had supposed, but a Virgo and that others who exercised the power of dowsing were born at all times of the day and night, in all seasons, and under every different phase of the moon. Jacques Aymar's brother, for example, born of the same parents, in the same month, under the same zodiacal sign, and in the same place two years after Aymar, did not have any dowsing capabilities whatsoever.[15]

Some doctors took a different tack in their analysis of the situation. The head of the medical college in Lyon, Jean-Baptiste Panthot, offered a physiological explanation, based largely on Aristotelian natural philosophy, to explain the talents exhibited by dowsers. He personally had witnessed Aymar in action and proposed that dowsers were somewhat akin to human magnets with the divining rod itself acting like the needle of a compass. In effect, the dowser could home in on the vapors left behind after an individual passed by. This would cause a physical, and frequently visible, reaction in the dowser. Physicians felt confident that this theory explained the violent headaches and fatigue Aymar occasionally suffered. He reportedly entered into a feverish state while practicing his art; he claimed he felt his body temperature increase, experienced muscle spasms and a quickening pulse rate, and, on occasion, he even vomited blood. The body of the dowser literally acted as a sponge into which the vaporous remnants of somebody's insensible perspiration could be absorbed. Medically speaking, dowsers traced individuals through a trail spread in their wakes with every breath they took. As a result, Aymar could not even walk near the ill-fated Joseph Arnoul without suffering severe heart spasms. Panthot and other doctors could not quite explain how Aymar could focus on one such individual, nor could they determine why Aymar could not trace non-criminals in a similar manner. It seemed that immoral behavior, leaving behind what Aymar called "murderous matter," was easier to trace, but nobody could invent a very good medical explanation as to why that might be. There were also various levels of intensity involved: Aymar claimed that he only fell violently ill when on the trail of particularly violent criminals and not when merely tracking thieves or finding springs and ores.[16]

Other participants in the debate theorized over the exact physical properties of the divining rod with an eye towards demonstrating that the mechanical philosophy could offer an explanation for Aymar's abilities. Savants such as Pierre Garnier, Pierre le Lorrain de Vallemont, and Pierre Chauvin, adopting a Cartesian point of view, suggested that people left behind them small, but very strongly constituted, corpuscles as they passed.[17] Cartesian explanations of light and heat might help us understand how the dowser made use of these corpuscles. From the Cartesian point of view, matter completely filled the world. The eye sees objects because they radiate corpuscles that bump into the matter between the object and the eye until the eye registers it. We can picture this as a sort of long solid line of matter from the object to the eye, an instrument designed specifically to filter such information. This sort of rationalization explained how heat could be transferred, for example from a candle flame to the hand. Cartesian natural philosophers expanded these examples to say that a dowser could read the matter left behind by certain individuals just as your hand would remain warm for some time after removing the source of the heat. Diviners used a tool, the divining rod, to focus these corpuscles, just as the eye focused and interpreted the light emanating from an object.

The initial attack against dowsing came most strongly from the religious sector. Theologians, as we might expect, had much to say on the topic of divining rods and the practice of divination in general. Apparently some individuals accepted the literal accuracy of the term "divining rod" and claimed that Aymar accomplished his feats through divine assistance.[18] But most religious explanations concentrated instead on an assumed diabolical intervention into the affairs of people on Earth. The Oratorian priest Pierre le Brun, for example, argued that Aymar succeeded in his work thanks solely to demonic assistance. Like Gassendi and Boyle, Le Brun did not dispute the fact that Aymar could find water, precious metals, or other material objects; in fact, he accepted the idea that physical objects had such scientifically based relationships as seen, most strongly, in magnets. However, when it came to Aymar's foray into the world of the soul and his seeming ability to identify moral and immoral activity with a divining rod, Le Brun felt that this clearly crossed the line between the realms of natural philosophy and demonic magic. He believed that in order for Aymar's material divining rod to move when it came close to the immaterial soul, the interference of an intelligent, intermediary agent played a vital role. And since Le Brun did not believe that God or angels would interfere in this way, demons remained the only other option. Le Brun assumed that the devil had somehow duped Aymar into accepting demonic help and, thus, had transformed the divining rod into a sort of magic wand. Unless the Church took steps, Le Brun lamented, this practice would surely spread even wider. In an age still replete with witches and sorcerers, these arguments were compelling, if not entirely original.[19]

In addition to the minimal trials performed initially by the authorities in Lyon, Aymar underwent several other experiments in an effort to verify and explain his abilities. In one instance, the Lieutenant-Général in Lyon, Matthieu de Sève, Baron de Fléchères, hid three *écus* under one of several hats on a table in his library and asked Aymar to find the money, something Aymar accomplished easily. He also asked Aymar to try to determine where, in his library, twenty-five écus had been stolen some seven or eight months earlier. Aymar first indicated the cabinet in which he had kept the money and then proceeded to trace the thief back to the servants' quarters and to the very bed the suspect had shared (and the side of it on which he usually slept), information corroborated by his former bedmate.[20] The wife of the Lieutenant-Général also expressed an interest in this affair and devised her own method of testing Aymar's abilities. Calling him into her drawing room, she asked Aymar to find out who had stolen money from a certain M. Puget, one of the witnesses present. It was a trick question, however, since she had taken the money herself. Aymar searched the room and announced that he did not believe a theft had occurred. She asked him to look again and he gave the same response as before but added, apparently rather coldly, that if there had been a theft it had been committed as a joke and in an innocent manner; his talent, he claimed, only worked when he tracked real criminals.[21]

These early experiments were supplemented in 1693, after much of the printed debate had taken place, by an attempt to answer once and for all whether dowsing held scientific validity. Henri-Jules, Prince de Condé, issued Aymar an invitation to come to Paris and submit himself to a series of tests. Condé assigned a member of his entourage, Monsieur Robert, to undertake this task and he, in turn, solicited the assistance of several members of the Académie Royale des Sciences and as well as some nobles who had expressed an interest in the proceedings. This gave the academicians in particular a chance to show their expertise and demonstrate their utility to the state at a time when some questioned the efficacy of the Académie. In one experiment, Aymar had to determine the amount and type of various metals buried in a garden. In another case, the academician Jean Gallois asked Aymar to find a gold Louis hidden in the garden of the Bibliothèque de Roi when, in fact, he had the coin hidden in his pocket. Robert asked Aymar to try to solve a theft at the residence of the Duchess of Hanover. Aymar indicated that the thief had left through the main door, but could achieve no further results. At Chantilly, the site of Condé's country estate, Aymar successfully identified the man who had stolen and eaten several trout from a basin in the Prince's gardens, but he also badly misidentified a boy as the man's accomplice. One of Aymar's detractors had lied and told Aymar the boy was the son of the guilty man when in reality the boy was not related to him at all and had been gone from Chantilly when the theft had taken place. Guided by misinformation, Aymar's divining rod had turned rapidly in the presence of the boy. Back in Paris, Condé gave Aymar one last chance. The experimenters led Aymar to the rue Saint-Denis, the site of an especially brutal murder of one of the king's archers who had come out short in an argument with some Musketeers. The man, reportedly stabbed fifteen or sixteen times, had bled profusely on the street. Aymar's divining rod did not twitch at all even though he passed over the exact spot of the murder several times. This test was flawed, Aymar claimed, both because the murderers had already been caught and because his talent only allowed him to detect premeditated crimes and not crimes of passion. Nonetheless, throughout all these experiments, both fair and foul, Aymar performed far below the expectations of Condé and the other witnesses. He returned to Dauphiné a humiliated man, and reports indicated that some time elapsed before he recovered all his dowsing skills.[22]

Several published statements appeared as a result of these experiments in an effort to warn the public of what Condé had discovered. Robert wrote a letter, published in journals such as the *Mercure Galant* and the *Journal des Savants*, claiming that Aymar's supposed talent was nothing more than an illusion and a trick. A few years later, the Académie Royale des Sciences put forward its official opinion in the form of a judgment passed on Le Brun's book against Aymar. This decision, published in 1701 and written by Bernard le Bovier de Fontenelle himself, but with the unanimous support of academicians such as Nicolas Malebranche, came out strongly against dowsing.[23] Aymar's failure before both the Prince de Condé and the academicians was a double blow to his claim for legitimacy and it would seem that the failed experiments held considerable weight in this matter, as least as far as certain members of the state were concerned. The evidence against Aymar notwithstanding, general support for dowsing did not waver. In other words, although nobles and academicians, as representatives of the state, came down on the anti-dowsing side of this debate, the overall consensus lined up in favor of dowsing. Why did popular support continue for Aymar and his work? The public stood behind Aymar largely because he continued to be successful, especially with regard to finding springs and ore deposits. He and other dowsers were so successful, in fact, that the French government largely ignored the decision of its specialists and nobles. Some people did not give up their belief even in the moral utility of dowsing. During the Revolt of the Camisards (1702–1705), for example, Aymar helped the Catholic side hunt down some Protestants accused of murder. He successfully completed this task and, at his word, a number of Protestants were executed. As late as 1706 Aymar appeared in Lyon to help local officials with a difficult case.[24] The crime-fighting side of dowsing did gradually die out, however, and by the end of the eighteenth century dowsers and their defenders would assert that their efforts were strictly scientific and that only charlatans claimed a moral utility for the science of rabdomancy.

This intellectual and cultural debate took place in several distinct public arenas. It occurred first, and most heatedly, in the pages of the *Mercure Galant* and the *Journal des Savants*. Oftentimes, several articles on dowsing appeared in a single issue. The debate continued into the next century in the pages of the *Journal de Trévoux* that offered judgments in the form of book reviews and broader commentary. Generally, the *Mercure Galant* allowed room for the debate to take full form, printing opinions on both sides of the debate and allowing for responses when appropriate. The *Journal de Trévoux* and the *Journal des Savants*, on the other hand, adopted distinct points of view. The former attacked dowsing from a religious perspective and the latter held that dowsing was certainly suspect from a legal and moral point of view and needed better theoretical justification from a scientific perspective as well. Many of these early commentators also published versions of their arguments in book form, several of which went through multiple editions that appeared as late as the 1750s and 1760s.[25]

The debate did not occur, however, merely in an abstract form. The experiments performed both in Lyon and at the behest of the Prince de Condé took place in noble homes (in Lyon, Paris, Versailles, and Chantilly), in private and public gardens, and in the streets of Paris where interested amateurs could look on. Participants in these trials included both men and women and, in cases such as that of the murdered archer, took the observers from the stately *hôtels* of the rich and famous down onto the bloodstained streets of Paris. The observers also ranged educationally with some savants taking an active role but, in other cases, also including keen amateurs with a ready interest but no real specialization or skill that gave them knowledge of the subject at hand. Thus, while most participants in this debate were scholars with some kind of official status, a few amateurs entered the debate, even if only as passive witnesses or as victims of the crimes Aymar struggled to solve.

Some of the debate occurred within the state system. This was less concrete in form, appearing instead as differences in state approval and sponsorship. The Académie Royale des Sciences, as state-sponsored arbiters of good science, took an anti-dowsing stance saying that scientific proof of its legitimacy did not exist. However, certain members of the royal administration essentially ignored this view and took advantage of dowsers whenever appropriate; a good waterfinder was still very useful. This continued to be true throughout the eighteenth century. Popular support for dowsing remained strong as long as dowsers continued to find springs and mineral deposits. Part of this was almost certainly a provincial versus Paris attitude with Parisians largely coming down against dowsing while those in the provinces, more of whom had seen it in action, backing it. In other ways this debate was about utility, mostly economic, and the level of incredulity found in the general population. Savants felt that the people who had witnessed Aymar's initial successes could not be trusted or taken at their word. Instead, participants in the debate elaborated a series of theories that either justified or discredited Aymar. When no single theory proved acceptable to the majority of savants, people simply fell back on the fact that dowsing, proven or not, had enormous utility. While some people believed in dowsing because it was so successful, it should also be noted that others might have backed Aymar because it was empowering to do so. At a time when the state tried to control all aspects of life, including science, the debate over dowsing allowed a group of people without state scientific authority to take a stance against the state. In some respects they even won the battle since certain parts of the government continued to allow and encourage dowsing for purposes of general utility.

Both the cultural and the state responses fit in rather nicely with literary historian Paul Hazard's theory of a crisis of consciousness at the end of the seventeenth century. Hazard even mentions the story of Aymar, amidst the vast number of other anecdotes he cites.[26] He claims that the ultimate rejection of Aymar's abilities by the state, after the turmoil of critical debate, reveals the general move in France towards a rational outlook that ends up in the age of Enlightenment. The broad mix of individuals and the breadth of cultural tools they brought to this debate certainly lend credence to Hazard's crisis of consciousness. However, the debate surrounding Aymar's feats of detection with the divining rod actually did not lead to as firm a conclusion as Hazard suggested. While the use of dowsing in legal proceedings clearly had limits, the question of whether dowsers could use divining rods to find water and minerals remained open. In other words, while critics could not prove or disprove the legitimacy of dowsing, the reality was that divining rods continued to work and so people continued to use them. As a result, Hazard's crisis of consciousness becomes somewhat dissipated. Without a resolution, the crisis evaporates and instead seems to exist as something deeper and more ingrained into the culture of the time.

It may be more useful to see this debate as a prototypical public sphere debate, one in which a limited number of doctors, natural philosophers, astrologers, theologians, nobles, amateurs, and savants all engaged in a discussion on dowsing and tried to influence public opinion. The state tried to take a stand, but ended up capitulating. The academicians expressed their view and people generally ignored them. Public opinion represented the dominant force in this case. Falling on the side of utility, public opinion triumphed over the views of diverse scholars who, whether supporting or attacking the practice of dowsing could never completely justify their positions and fully persuade the public at large. The Enlightenment begins, then, not with the victory of rationality over superstition but with the creation of a public forum in which the relative positions of rationality and superstition could undergo open debate and discussion. The goal of the individuals involved in the debate was not to decide if Aymar really could solve crimes but to assert authority; and when, in the end, people essentially ignored the authority of savants, nobles, state representatives, and theologians in favor of utility, it becomes clear that they had defied the attempts to sway their opinions. In fact, the general population that had observed the successes of dowsing simply continued using it as they always had. It was the scientifically unfounded opinions of the popular classes that carried the day and it was this point of view that continued into the eighteenth century.

Enlightened dowsing and Barthelemy Bléton

Several dowsers appeared during the late seventeenth and eighteenth centuries, allowing some continuity to the debate over its validity. Two young women, Mesdemoiselles Ollivet and Martin, obtained notoriety at about the same time as Aymar. Ollivet felt deeply troubled by her ability to use a divining rod to find water and so consulted Pierre le Brun for help. He advocated prayer. After two days spent in solitary contemplation Ollivet underwent several tests supervised by Le Brun and was pleased to find that God apparently had vanquished her dowsing capabilities. Martin, the daughter of a merchant from Grenoble, seemed less bothered by her talent; she enjoyed the gift of finding and determining the authenticity of relics.[27] Later in the eighteenth century a human "hydroscope" appeared on the European dowsing scene. Jean-Jacques Parangue could apparently "see" water hidden underground in much the same way we can see vapors rising from the street on a hot day. According to one explanation, he had "lynx eyes." The Abbé Sauri, who wrote a short treatise on dowsing and associated it with ventriloquism, based his theories on the number of fibers in Parangue's retinas and assumed that as the young man aged, his eyesight would weaken and he would lose his remarkable ability. This episode in the history of dowsing sparked a scientific parody. An anonymous author published the mock history of a young English woman named Jenny Lesley who shared Parangue's talent.[28]

During the eighteenth century the nature of the debate over dowsing underwent a metamorphosis. Voltaire, with his usual panache, called dowsing bad science and suggested that the divining rod should be banned from France along with astrology, magic, demonic possession, the cabala, theology, and Jesuits.[29] Generally speaking, analyses of dowsing came to concentrate more on whether it was good or bad science than on its possible religious or magical qualities; in other words, the debate was rationalized. Aymar, for example, appears in the *Encyclopédie*, but largely as a negative example of how dowsing should not be practiced. The article on the divining rod (*baguette divinatoire*), penned by Jean d'Alembert, is a case in point. D'Alembert asserted that Aymar had been, without doubt, deceitful and duplicitous. He does not, however, suggest that there was anything magical or demonic about Aymar's ability to solve crime. As for finding water and the other, more natural, aspects of dowsing, d'Alembert did not completely dismiss them but noted that they had been called into question. The article on *rabdomancie* offered an extended historical look at dowsing before getting to the case of Aymar. Once again, while the author portrayed the use of divining rods to solve crimes as silly, the question of their potential scientific merit appeared less obvious. In other words, just because Aymar behaved duplicitously does not mean that all dowsers were frauds.[30]

Later, during the 1770s, the debate picked up steam again with the appearance of Dauphiné's other famed dowser, Barthelemy Bléton. Born sometime during the 1740s into a peasant family, Bléton discovered his abilities at the age of seven. While carrying dinner to some workmen, Bléton took a break and sat down on a large rock where a sudden fit came over him; he was faint and feverish and did not feel any better until he was moved from that spot. Every time he went anywhere near the stone, his illness returned. After witnessing Bléton go into his fit, the local prior concluded that there must be something about that specific location that caused the sickness and so ordered some men to dig up the ground around the big rock. There they found a spring which, according to one source, was still in use at the beginning of the twentieth century.[31] Since the region of Dauphiné had already produced a number of famous dowsers, Aymar among them, the people there were always on the lookout for this type of occurrence. This time, however, the debate over dowsing focused largely on issues of scientific authority and legitimacy. The chief question centered on who had the right to verify something as scientific. The pro-dowsing faction received backing from a few savants and from several nobles, most notably Louis XVI and Marie-Antoinette who used Bléton's talents both for their own purposes and for the general benefit of France. But Bléton received his most powerful support from the general public itself. Members of the French population at large continued to hire him, witness his feats, and write testimonials concerning his success.

By the time Bléton arrived on the scene, the dynamics of the debate had changed. At that point the controversy fully entered the public sphere. In the previous case of Jacques Aymar a relatively small group participated in the discussion, including a number of people and institutions (like the emerging periodical press). But by the time Bléton gained public attention, the entire range of the argument had entered the domain of public opinion. Amateurs, keen to apply their understanding of Enlightenment reason and convinced of their ability to grasp the laws of nature, began to take sides and offer opinions. Savants were unhappy with this trend and declared that the Parisian public should not involve itself in scientific debates. The boulevards were not a place, they claimed, for the testing of theories.

Some savants perceived the credulity of the general population as a growing problem in late eighteenth-century France partially due to the success of scientific popularization in reaching a growing portion of the urban population. In the decades before the French Revolution, a wide variety of activities existed that disseminated scientific ideas to willing Parisians. Public lecture courses, along with popular demonstrations and scientific clubs, all spread science to an eager and increasingly large audience. In this environment, Parisians rapidly appropriated science into the general urban culture. As a result, some people began to feel confident in their ability to participate in current scientific discussions and savants occasionally turned to the public sphere for support. In the 1770s and 1780s several scientific battles were being waged in the public sphere.[32] Two controversies in particular, over medical electricity and Franz Anton Mesmer's animal magnetism, were closely tied to that of Bléton and his divining rod. All three practices depended on some sort of *baguette* or wand (to apply electricity to a specific spot, aid the flow of animal magnetism, and, in the case of the divining rod, to locate ores or water in a particular region) and they all relied on invisible and frequently inexplicable fluids as explanations.[33]

Bléton received his most active individual support from a minor savant named Pierre Thouvenel. Born in 1747, Thouvenel received a medical degree and established himself in Paris where he soon became identified with questions of water supply, and received an appointment as the royal inspector of mineral waters in France. It was through his work in this capacity that he first came into contact with Bléton. Ultimately, Thouvenel wrote several books and innumerable letters and articles arguing for the validity of dowsing. He published his first book on the subject, the *Mémoire physique et médicinal*, in 1781. In this volume, Thouvenel tried to present a clear account of Bléton's abilities and justify dowsing's place as a science. Linking his own work, and oppression, to that of Galileo, Thouvenel set himself up as an unsung hero of scientific discovery who fought to bring the radically new science of rabdomancy to its deserved central position within the scientific world.[34]

In the book's first section, Thouvenel established general physical propositions he thought might help explain scientifically how Bléton and other dowsers performed their work. Specifically, Thouvenel based dowsing on a combination of electricity and magnetism, a sort of "animal electricity." He claimed that certain individuals were more sensitive to changes in the fluids that surround and flow through us all and, therefore, could sense minute alterations in the underground flow of electricity and magnetism.[35] The changes in electric fluids accounted for the fits associated with dowsing and the appearance of seizures akin to epilepsy. The seizures, like the medical crises associated with mesmerism, supported a connection between magnetism and the ability to discover ore deposits. Although Thouvenel wanted to turn dowsing into a scientific subdiscipline of physics, he was largely unable to prove the connections he suggested. This meant that he had to supplement his physical and medical conjectures with additional materials.

Thouvenel offered this further proof in the second section of his book where he described, in copious detail, his own observations on Bléton that convinced him the dowser could in fact find underwater springs and minerals. Here we get a meticulous account of the spasms and convulsions that shook Bléton while he worked. Thouvenel also examined the motion of the divining rod which, when near water or a mineral deposit, would begin spinning at an estimated speed of thirty to eighty rotations per minute. Notably, Bléton did not utilize the more traditional forked branch but used a single, slightly curved piece of wood instead. In addition, he claimed that he only used the divining rod for the benefit of observers; he himself could actually feel the presence of water beneath his feet and did not need the twitching or rotating rod to help him. In this section, then, Thouvenel described Bléton and the practice of dowsing as witnessed through countless demonstrations. In grand fashion, Thouvenel did not want to feign any hypotheses and so spent a great deal of time observing the practice of dowsing. However, while he presented these observations as experimental in nature, they were really just descriptions of Bléton at work and, as such, not easily disproved by savants who sought to attack dowsing. Simply put, there was no concrete experimental procedure to examine and refute.

The third and last part of Thouvenel's book neatly took the science out of this debate and put the question into the hands of the general public. This section reproduced numerous notarized documents, reports, and affidavits that attested to the success Bléton had enjoyed throughout France. For the most part, these were testimonials from satisfied clients including clergymen, nobles, town councils, and members of the professional and middle classes. It was much harder to refute these documents than it was to disprove Thouvenel's theories or observations even though they did not technically have any scientific value. They were, after all, merely testimonials from amateur observers and, as such, carried no scientific weight. They simply represented the vast number of individuals who claimed to have watched Bléton successfully find springs and mineral deposits. The problem for critics lay in the status of these witnesses: were they able to be good witnesses or were they too credulous of things they did not actually understand, despite their education, or even noble or clerical status?[36]

The reviewer of Thouvenel's book for the *Journal des Savants* recognized the ambiguous position of dowsing and tried his best to find a middle ground. He wanted to "hold himself equally distant from the blind credulity of many of the ignorant and from the oftentimes too presumptuous incredulity of certain savants." The resulting review was necessarily vague; Thouvenel's book clearly lacked scientific proof, but the obvious utility of dowsing and its apparent success rate were difficult to deny. The problem for this reviewer, and for many opponents of dowsing, lay in the fact that Thouvenel claimed scientific status for dowsing, but supported his theories through public acclamation rather than mathematical or experimental proof. Not that everybody suffered the same reticence about attacking Thouvenel. Louis Bernard Guyton de Morveau criticized Thouvenel for, among other things, not including the full names of those who had testified to Bléton's talents. He insinuated that perhaps Thouvenel had written all the documents himself. The mathematician Gabriel Antoine de Lorthe critiqued Thouvenel on a more basic level. He noted that Thouvenel had forgotten to include any physics and medicine in his book and suggested that the tome had been misnamed – Lorthe recommended a new title, the "Story of the Celebrated Bléton." J.P. Battelier, in the *Nouvelles de la république des lettres et des arts*, expressed his amazement that so many people, commendable by their lineage and knowledge, should have allowed themselves to be convinced of the reality of dowsing when it was so clearly "opposed to reason, [and] even to nature."[37]

Thouvenel's book inspired a virtual cacophony of opinions as savants and amateurs alike debated the acceptability of dowsing. On one side Bléton and Thouvenel were praised for seeking the truth and trying to make known, for the benefit of the nation as a whole, a new scientific discovery. Several savants, for example the well-known demonstrator of experimental physics Joseph Aignan Sigaud de la Fond, spoke in favor of dowsing. Sigaud de la Fond discussed the practice in his *Dictionnaire des merveilles de la nature*, a work in which he described many seemingly inexplicable but nonetheless, in his opinion, scientific phenomena. Sigaud de la Fond explicitly praised the utility of the divining rod: it could be used to discover, he claimed, "springs, mines, and the hidden treasures of the earth." Unfortunately, he added, charlatans who took advantage of "public credulity" could also abuse the divining rod.[38] As an example, he pointed back to the case of Jacques Aymar who, according to Sigaud de la Fond, could never have used the scientifically sound divining rod to solve a murder.

Dowsing's opponents did not mince words in their attack; they labeled Bléton a sorcerer [*sorcier*], a pun on the French word for a dowser, *sourcier*.[39] They directed much of their attention to the motion of the divining rod itself, or what the pro-dowsing group had termed the hydroscope. This was the most easily controlled and mimicked part of the dowsing process and the one most susceptible to the skill of a charlatan. The esteemed astronomer and academician Joseph-Jérôme Lalande, for example, published a letter in the *Journal des Savants* that purported to prove how Bléton moved his divining rod. Another detractor, Henri Decremps, devised an experiment using the divining rod and then gave detailed, step-by-step instructions on how any competent charlatan could reproduce it and convince people that dowsing really worked.[40]

Unfortunately for Lalande, Decremps, and others opposed to dowsing, showing how someone could purposefully manipulate a divining rod to rotate and vibrate did not prove anything. No one had ascertained that Bléton practiced false methods. In fact, as noted earlier, Bléton claimed he did not even need the divining rod; his entire body gave evidence of the existence of underground springs. Furthermore, showing how to perform a parlor trick with a divining rod did not explain Bléton's amazing success rate in finding springs. Despite the negative endorsement of several well-known savants published in a variety of periodicals, dowsing maintained its public credibility. J.P. Battelier pointed out that these savants were generally recognized as witnesses of integrity and worth. Nonetheless, their opinions were ignored, something that Battelier blamed on the enthusiasm evinced by some periodicals. As a result, he claimed, public opinion fell almost unanimously with Bléton.[41] Clearly, the opposition needed to debunk Bléton in a way that allowed public participation in the experimental process. In other words, the public needed to actually witness his failure. Thus, the anti-dowsing group moved to address the public on their own ground. To achieve this end, they invited Bléton to visit Paris to participate in experiments, before a general audience, designed to prove once and for all that dowsing was not scientifically sound.

Thouvenel and Bléton did indeed come to Paris, but the first round of experiments did not go exactly as the anti-dowsing faction would have liked and public support for this useful but inexact science remained high. Early in May of 1782 Bléton went to the Jardin de Luxembourg and submitted himself to several days of public scrutiny. Charles-Alexandre Guillaumot, the Intendant-Général des Bâtiments du Roi, presided over the experiments, accompanied by numerous doctors, savants, men of letters, artisans, and distinguished amateurs – in all, over 1200 people attended the demonstrations. The main part of the experiment called for Bléton to try to trace an underground aqueduct, the plan of which was in the sole possession of Guillaumot. Bléton met with nothing but success. In fact, Guillaumot claimed that Bléton's path above ground so accurately mapped the underground flow of water, that if the plans for the aqueduct were ever lost, they could easily use Bléton to determine its exact location. He could even determine how deep the aqueduct ran and the diameter of the pipe being used. Bléton performed this feat not once but twice, on two different days and, to add insult to injury, did it blindfolded both times. The *Journal de Paris* announced that the science of dowsing could now be appropriated for the benefit of "physics and for the economic utility of society."[42] Once again, Bléton had publicly demonstrated his talents and the anti-dowsing side was left wondering where they had gone wrong. Clearly, their effort to debunk dowsing was going to require more rigorous experiments along with a larger appeal to the sensibilities of the general public.

While the anti-dowsing group scurried to create a new set of experiments, the pro-dowsing faction used their victory to strengthen their cause. Bléton kept himself busy in and around Paris amazing and astounding royal ministers and officials, ambassadors, clergymen, and countless other amateurs as he continued to find springs. While visiting the Jardin de Pharmacie, ostensibly to attend a public lecture, he agreed to submit himself to an experiment at the hands of some of the professors. The *Journal de Paris* reported that from 1:00 p.m. until 6:00 p.m. Bléton endured considerable pain in order to determine the connection between his powers and electricity. Essentially, the members of the Collège wanted to know if Bléton could distinguish the rate of flow of electricity just as he could determine the rate of flow for water. After submitting Bléton to five hours of electric shocks, the answer was yes; nine members of the Collège, along with two visiting physicists, signed a letter to that effect.[43] Bléton won converts almost as rapidly as he found springs.

But by the end of May, the anti-dowsing faction was ready to try again. On 29 May Bléton showed up at the garden of the Abbaye de Sainte-Geneviève to undergo testing before a huge audience. In addition to the large number of amateurs, the tests were witnessed by members of the Académie Royale des Sciences including Bertrand Pelletier, J.-A.-N. Caritat, marquis de Condorcet, and Charles Bossut; other luminaries such as Benjamin Franklin, Denis Diderot, and the baron d'Holbach; and several nobles and state dignitaries. At the same time, there were a number of scientific popularizers present, such as Nicolas-Philippe Ledru. He and other mid-level savants worked full-time practicing science, but earned their income and their legitimacy through public rather than state support. Success in their world depended on drawing a large, amateur clientele for their popular science classes. By including them as witnesses to this series of experiments, the anti-dowsing side may have hoped to sway public opinion to their side and detract from the affidavits printed in Thouvenel's book.

This set of experiments certainly made a pretense at being more rigorously developed and more closely watched. The garden, already replete with fountains and running water, was prepared in advance with hunks of minerals and ores buried underground at different depths and in various amounts. Bléton, with his eyes covered, zigzagged back and forth across the garden for eight hours before his blindfold was removed. Throughout the day he "found" many ores and water sources but, unbeknownst to him, he also missed many. A week later Bléton returned to the garden. This time the anti-dowsing side wanted to see if he would indicate the same spots he had found the week before. Also, they wanted to examine more closely the movement of the divining rod. Bléton arrived in the morning, blindfolded as before, and again roamed the gardens indicating springs and ore deposits until half past noon; but he had not been able to duplicate his feat from the previous week. At the same time, no mathematical correlation between the rotation of the divining rod and his proximity to either underground water or ores could be identified. These results, printed in full in the *Journal de Physique*, were also published in summary form in the *Journal de Paris* in order to reach the widest possible audience.[44]

This, of course, did not end the matter. Within two weeks Thouvenel, who claimed that the tests performed by the anti-dowsing faction were invalid, took Bléton back to the gardens to perform some tests of his own. He published his results, along with some additional affidavits, in a lengthy supplement to the *Journal de Paris*. He also continued to solicit support from several savants such as Pierre-Isaac Poissonnier, Pierre-Joseph Macquer, Jean Darcet, and Joseph-Ignace Guillotin.[45] In the meantime, Bléton continued to find springs in and around Paris itself, often at the request of the Intendant of Paris or other royal officials. The anti-dowsing faction had succeeded in demonstrating before a crowd of savants and amateurs that Bléton was inconsistent at best, but they still were unable to explain how Bléton seemed to find more springs than anyone else. Nonetheless, public opinion began to waver a bit; Bléton's failure to live up to the standards set by the academicians, at least when he failed to meet those standards in a public forum, did hold some stigma.

At this time the popularizer, and soon-to-be famous aeronaut, Jacques-Alexandre-César Charles threw his hat into the ring on the side of the anti-dowsing group. Bléton agreed to undergo some experiments designed by Charles who wished to examine the connections between electricity and dowsing. Bléton stood over an aqueduct where, of course, his divining rod began to move. Then he stood up on an insulated stool. Sure enough, the rod stopped its motion. This was repeated several times but in the final experiment, Charles secretly connected a wire from the ground to Bléton. The divining rod, however, did not move. This led Charles, and many others, to conclude that the rotation of the rod was entirely the doing of Bléton.[46] Charles enjoyed a solid reputation among the general public and his experiments were probably more damaging to Bléton's image than those performed by the academicians.

Thouvenel and Bléton were now on the defensive as popular opinion began to shift. At about this time, Mesmer was also coming under attack.

Independently, Bléton could probably have survived for some time as long as he kept finding springs. But with the attacks against animal magnetism coming fast and furiously, and with the obvious connections between mesmerism and dowsing, the pro-dowsing side faced growing hurdles. Popular opinion against animal magnetism negatively affected the position of dowsing; it did not help their cause any that the mesmerists themselves explicitly associated animal magnetism with dowsing, citing Thouvenel and Bléton as allies.[47] The battle raged for the next several years. The anti-dowsing side, claiming for themselves alone the "love of truth," continued to write letters against Bléton that appeared in popular periodicals. The *Journal de Paris*, as the unofficial news agency of the pro-dowsing faction, continued to print articles citing Bléton's successes and letters from provincial subscribers who had witnessed his abilities. The ballooning craze, beginning in 1783, worked in Thouvenel's favor by making the impossible and seemingly inexplicable seem reasonable and commonplace. After the Montgolfier brothers and their imitators took to the skies, Thouvenel wrote that Bléton was to the subterranean world what the aeronauts were to the heavens.[48]

The issue ultimately remained unresolved as can be seen from the number of nineteenth- and even twentieth-century books that either explore the public benefits of dowsing or attack it as a fraud.[49] For their part, Bléton and Thouvenel could never mathematically prove their theories. On the other hand, while the anti-dowsing side showed how a charlatan could turn the divining rod, they could never explain how Bléton continued to find springs and mines with remarkable frequency. When the second volume of Thouvenel's *Mémoire physique et médicinal* was published in 1784 it received the same sort of ambiguous attention by savants as his first book. The reviewer for the *Journal des Savants* merely noted that this volume contained the same kind of information that Thouvenel had been publishing in the *Journal de Paris*, but that the marvels were even more numerous in the new book. He did not directly criticize Thouvenel, Bléton, or the practice of dowsing although he did mention that Thouvenel's theories remained unproven. His reluctance to attack the pro-dowsing side may have been caused by Bléton's discovery at about that time of a particularly large and lucrative coalfield.[50] Public opinion, charmed once again by this concrete evidence of utility, had temporarily resumed its active support of Bléton. Subsequently, Bléton managed to maintain his somewhat battered reputation and continued to work with Thouvenel finding springs and ore deposits for the state. In the first six months of 1785 he discovered forty-one coalfields in ten different provinces throughout France.[51]

The debate might have continued occupying the public mind had not other concerns arisen to distract the people of France from dowsing. Scientific wonders like ballooning, polemical debates over mesmerism, and, of course, the onslaught of the French Revolution turned popular attention away from Thouvenel and Bléton and on to more pressing issues. Utility would sometimes win out over scientific opinion and the public still occasionally rose to defend its practice; ultimately, however, dowsing no longer had enough support to claim scientific status. Thouvenel suffered disgrace within the scientific community and no amount of public support could help him sway savants to accept his suppositions as factual. He eventually moved to Italy, taking with him another dowsing protégé from Dauphiné named Pennet, where he lived out his days trying to prove his theories.

The debate over Bléton covered much more ground than the one concerning Aymar. First, the sites of the debate were greater and reached more people. Found in publications like the *Journal de Paris*, the *Journal Encyclopédique*, and the more scholarly *Journal de Physique*, the struggle to prove dowsing's legitimacy reached a much larger audience than previously had been possible. In addition, the growth in literacy and the greater circulation of these journals, especially the *Journal de Paris*, ensured that a much larger portion of the population followed the currents of this struggle more easily than ever before. At the same time, and in this way similar to Aymar, the debate over Bléton did not occur solely in print. Bléton performed numerous demonstrations, all attested to and verified by members of the general public. In his case, these demonstrations of his abilities were made all the more potent thanks to Thouvenel's publication of the affidavits in his books and in the *Journal de Paris*. In this way, dozens of non-specialists could assert their authority as witnesses and vouch for Bléton's skills. Similarly, the very open and very large public demonstrations around Paris drew huge crowds of people. Authority, in this case, lay with public opinion and, more importantly, with whoever could sway public opinion. Much to their chagrin, savants found that in order to do that they had to abandon their theories and ideas and focus instead on the visual, the observable, the amusing, and the useful. High science had to meet the popular practice of dowsing on its own ground, within an open arena.

The increasing availability of popular science and its increasing appropriation by the public helped create a new place for natural philosophy in society. Simultaneously, the tenor of the debate over divining rods changed. In the case of Bléton, gone were the appeals to astrology and theology that appeared at the time of Aymar. Guyton de Morveau claimed, for example, that Bléton should not be accused of magic but of being a swindler.[52] The question had shifted to one of good use of reason versus popular credulity rather than one of demonic or astral influences battling against Cartesian corpuscles. The size of the audience had clearly grown too. With Aymar, a small group, distinctive by the nature of their education, debated among themselves and sought to influence their intellectual peers. By the time Bléton appeared on the scene, little or no control existed over who could read, discuss, or contribute to the debate. Academicians, amateurs, and popularizers all spoke their minds and sought to influence the outcome of the debate.

Members of the general public felt confident in their ability to participate in scientific discussions. A half-century spent appropriating various types of science and obtaining a substantial level of cultural scientific capital left many members of the public convinced that they were equally able to judge scientific affairs. The availability of popular science sharply increased the number of people willing to assert their own scientific opinions. The danger, from the point of view of elite savants, lay in the fact that not all versions of popular science were equal and while many felt ready to face the challenge of deciding scientific debates, not everyone enjoyed the same level of preparation. In giving people an awareness of science, popularizers sometimes imbued them with a greater sense of knowledge than they actually possessed. Popular scientific knowledge, then, could be used for purposes far beyond its original intent and the level at which it had been appropriated.

The debate over dowsing occurred within a public forum and even though the issue was not resolved, it took place with the state, savants, and public all participating as they saw fit. If there was anything radical about the debate over dowsing, it was that the state, savants, and public all participated on relatively equal terms. The Enlightenment, at least those aspects of it appropriated by a general public fascinated by dowsing, was out of the control of the elite few and into the hands of the many.[53] The public took over the task of creating their own opinions, temporarily, and passed judgment over the rationality and utility of a scientific question. The phenomenon of public engagement with scientific matters reached fruition with the invention of balloons.

Notes

1 A billhook is a tool with a curved blade attached to a handle, used especially for clearing brush and rough pruning. Unfortunately, I have been unable to discover the name of the wine merchant's wife.

2 Throughout the time Aymar tracked the culprits, he repeatedly duplicated this feat and thoroughly amazed his police escort and other spectators with his ability to identify the beds, chairs, plates, knives, bottles, and glasses used by the suspects. See the account by the Abbé de Lagarde, "Histoire du fait," in Pierre Garnier, *Histoire de la baguette de Jacques Aymar* (Paris: Jean-Baptiste Langlois, 1693), 80–2; and Pierre Viollet, *Traité en forme de lettre contre la nouvelle rhabdomancie* (Lyons: H. Baritel, 1694), 12.

3 There are several contemporary summaries of these events: see the eyewitness accounts of Jean-Baptiste Panthot in his *Lettre de M. Panthot* (Grenoble, 1692) and his *Traité de la Baguette*, 3rd edn (Lyon: Thomas Amaulri & Jacques Guerrier, 1693); Garnier, *Histoire de la baguette*; and [Jean] Vagini, "Récit de ce que Jacques Aymar a fait pour la découverte du meurtrier de Lyon," in Pierre le Lorrain de Vallemont, *La Physique occulte, ou Traité de la baguette divinatoire*, 2 vols. (The Hague: Moetgens, 1747), I:I:29–49, of which there is also a 1693 single-volume edn. For modern synopses see Alphonse Gilardin, "Un procès à Lyon en 1692, ou Aymar, l'homme à la baguette," *Revue du Lyonnais* (1837): 81–99; Louis Figuier, *Histoire du merveilleux dans les temps modernes*, 4 vols. (Paris: Hachette, 1860), II:59–70; William Barrett

and Theodore Besterman, *The Divining Rod* (Toronto: Coles, 1979), 27–31; and Paul J. Morman, "Rationalism and the Occult: The 1692 Case of Jacques Aymar, Dowser *Par Excellence*," *Journal of Popular Culture* 19 (1986): 119–29.

4 For a general history of dowsing see Barrett and Besterman, *The Divining Rod*, including their excellent bibliography of primary source material; and Besterman, *Water-Divining: New Facts and Theories* (London: Methuen, 1938). Most modern works on dowsing do a poor job of providing a scientific justification for the practice. However, for an exhaustive case analysis done by a physicist (who had started out trying to debunk dowsing) see Barrett, "On the So-Called Divining Rod, or Virgula Divina," *Journal for Psychical Research* 13 (1898): 2–282 and 15 (1900): 130–383.

5 In fact, Robert Darnton briefly cited the case of Bléton, among examples of ballooning and elastic shoes that allowed the wearer to walk on water, in his book *Mesmerism and the End of the Enlightenment in France*, to illustrate the variety of scientific theories that confronted Parisians at the same time as those of Franz-Anton Mesmer. Darnton, *Mesmerism* (Cambridge, MA: Harvard University Press, 1968), 31 and 96 (Bléton), 23–4 (elastic shoes), 18–22 (balloons).

6 Pierre Jean Claude Mauduyt de la Varenne, "Nouvelles Médicinales et Physique." In *La Nature considérée sous ses différens aspects*, ed. Pierre-Joseph Buchoz, 5 vols. (Paris, 1780–1783), IV:82.

7 In general, see Mona Ozouf, "'Public Opinion' at the End of the Old Regime," *Journal of Modern History* 60 (1988, supplement): 1–21; on medicine see L.W.B. Brockliss and Colin Jones, *The Medical World of Early Modern France* (Oxford: Clarendon, 1997), 590–605; on art see Thomas Crow, *Painters and Public Life in Eighteenth-Century Paris* (New Haven, Yale University Press, 1985), chp. 7.

8 Martine de Bertereau, baronne de Beausoleil, *Véritable déclaration de la découverte des mines et minières de France* (N.p., 1632) and her *La Restitution de Pluton* (Paris: H. du Mesnil, 1640). A description of these works can be found in Figuier, *Histoire du merveilleux*, 276–311. According to Figuier, Bertereau was also the first person to mention that divining rods could be used for the discovery of underground waters, the use for which they are most known today: see Figuier, 312.

9 Lynn Thorndike, *A History of Magic and Experimental Science*, 8 vols. (New York: Columbia University Press, 1923–1958), VII: 256 (Agricola), VIII: 495 (Gassendi and Boyle).

10 Pierre Bayle, *Dictionnaire historique et critique*, 16 vols. (Geneva: Slatkine, 1969), I:9. See also Viollet, *Traité en forme de lettre*, 4 (for adultery) and 163 (for virginity); *Mercure galant*, September 1693, 235 (virginity); and *Affiches de Dauphiné*, 6 November 1778, 111 (abandoned children).

11 Claude Comiers, *La Baguette justifée, et ses effets démontrez naturels* (N.p., 1693), 26–7.

12 Almost all contemporary accounts of Aymar give some historical background. Generally, those in favor of dowsing cite more biblical examples while those against it cite pagan examples. For a pro-dowsing example see Comiers, *La Baguette justifée*, 51–3; for an anti-dowsing example, and probably the longest historical look, see Pierre le Brun, *Histoire critique des pratiques superstitieuses* (Paris: Jean de Nully, 1702), 73–173.

13 A nativity is a horoscope based on the date of a person's birth. See the astrological chart cast for Aymar reprinted in the 1747 edition of Vallemont, *La Physique occulte*, II:165.

14 For the astrological influences involved in the cutting of the divining rod (accompanied by illustrations of sample horoscopes), see Vallemont, *La Physique occulte* (Amsterdam: Adrian Braakman, 1693), 386–94; also see Jean Nicoles, *La Baguette divinatoire, ou Verge de Jacob*, ed. Paul Chacornac (orig. edn 1693; Paris: La Diffusion Scientifique, 1959), 117–18. Nicoles' book has been translated as *Jacob's Rod*, trans. Thomas Welton (London: Thomas Welton, 1870).

15 Claude-François Ménestrier, "Des indications de la baguette pour découvrir les sources d'Eau, les Métaux cachez, les Vols, les Bornes déplacées, les Assassinats, &c.," in *La Philosophie des images énigmatiques* (Lyon: Hilaire Baritel, 1694), 442–3; Thorndike, *A History of*

Magic and Experimental Science, VIII:500. On Aymar's brother see Comiers, *Baguette justifiée*, 25. Clearly, not all astrologers were in accord as some emphasized the individual's birth while others suggested that the key to rabdomancy centered on when the dowser cut his or her rod.

16 Panthot, *Lettre de M. Panthot*, 8–10. On the response of the medical community to Aymar's abilities see Come Ferran, "Les médecins de Lyon et la baguette divinatoire au XVIIe siècle," *Bulletin de la Société Française d'histoire de la médecine* 30 (1936): 225–43. Of course, some doctors came out against Aymar, but usually for theological, and not medical, reasons.

17 Garnier, *Histoire de la baguette*; Le Lorrain de Vallemont, *La Physique occulte* (1747); and Chauvin, *Lettre à Madame la marquise de Senozan* (Lyon: J.-B. de Ville, 1693). See also Comiers, *Factum pour la baguette divinatoire* (N.p., 1693).

18 Lagarde, "Histoire du Fait," in Garnier, *Histoire de la baguette*, 88.

19 Pierre le Brun, *Histoire critique des pratiques superstitieuses* (Paris: Jean de Nully, 1702), 173–83. Pierre Viollet also believed that demonic assistance was required in the moral use of divining rods, but, like Le Brun, suspended judgment on their use in finding water and minerals. See Viollet, *Traité en forme de lettre*, 4–5. Also see A. Boissier, *Recueil de lettres au sujet des maléfices et du sortilège* (Paris: Charles Osmont, 1731), 62–9.

20 The suspected servant could not be questioned as he had subsequently left the service of the Lieutenant-Général.

21 Garnier, *Histoire de la baguette*, 101–2.

22 Condé gives no reason why he decided to invite Aymar to Paris and put him to the test. For a summary of some of these experiments see Figuier, *Histoire du merveilleux*, II:79–88; Jacques Barthelemey Salgues, *Des erreurs et des préjugés répandus dans la société* (Paris: Buisson, 1810–1811), 136–40. On the tests given by Condé see Paul Bussière, *Lettre à M. l'abbé D.L.**** (Paris: Louis Lucas, 1694), 18–23 (for the trout), 39–40 (for the archer).

23 Le Brun, *Histoire critique des pratiques superstitieuses*, 2nd edn, 4 vols. in 2 tomes (Amsterdam: Jean-Frederic Bernard, 1733–1736), I:lx; reprinted in Michel-Eugène Chevreul, *De la baguette divinatoire* (Paris: Mallet-Bachelier, 1854), 107–8.

24 The Intendant in Burgundy, for example, reported in 1707 that a dowser had discovered a mine up in the nearby mountains. "M. Pinon, intendant en Bourgogne, à M. Desmaretz," 21 Juillet et 24 Novembre 1707, in *Correspondance des Contrôleurs généraux des finances avec les intendants des provinces*, 3 vols. (Paris: Imprimerie Nationale, 1883), II:423. On the Camisards, see Jean-Baptiste Louvreleuil, *Le Fanatisme renouvelé*, 3rd edn, orig. edn, 1704, 3 volumes in 1 tome (Avignon: Seguin, 1868), II:51–2; and Antoine Court, *Histoire des Troubles des Devennes, ou De la guerre des Camisards*, 3 vols. (Alais: J. Martin, 1819), I:353–6. It is unclear if Aymar really helped in the War of the Camisards; the entire incident may have been fabricated in order to emphasize the intense superstition of one of the Camisard leaders. On the 1706 case in Lyon see Nicolas Boileau-Despréaux, *Correspondance entre Boileau-Despréaux et Brosette* (Paris: Techener, 1858), 267.

25 The two most popular books on the case of Aymar, those of Le Brun and Le Lorrain de Vallemont, went through half a dozen editions each appearing into the 1760s. On public opinion, the Republic of Letters, and the periodical press see Anne Goldgar, *Impolite Learning: Conduct and Community in the Republic of Letters* (New Haven: Yale University Press, 1995), esp. chp. 2.

26 Hazard, *The European Mind: The Critical Years, 1680–1715*, original edn 1935 (New York: Fordham University Press, 1990), 177–9.

27 On Ollivet and Martin, whose first names have not survived, see Pierre Le Brun, *Histoire critique des pratiques superstitieuses*, 4 vols., 2nd edn (Paris: Delaulne, 1732–1737), III: 374–88; and Bernard Forest de Belidor, *Architecture hydraulique*, 4 vols. (Paris: Charles-Antoine Jombert, 1737–1753), II:343–4.

28 On Parangue see the Abbé Sauri, *L'Hydroscope, et le ventriloque* (Amsterdam; Paris: Valade, 1772), 39–40. On Jenny Lesley see, *Histoire . . . d'une jeune angloise* (Paris: Lottin, 1772). For contemporary response to Parangue and Lesley see *Correspondance littéraire, philosophique*

et critique, ed. Maurice Tourneux, 16 vols. (Paris: Garnier, 1877–1882), X:43–4; *Année littéraire* 19 (1772:5):49–56; and *Avant-Coureur*, 15 June 1772, 374–7; ibid., 29 June 1772, 407–11; and ibid., 7 September 1772, 573–4.

29 Voltaire, *Lettres à S.A. Mgr. le Prince de ******, in *Oeuvres complètes de Voltaire*, ed. Condorcet et al. (Paris: Garnier, 1879), *Mélanges*, V:488.

30 All references from the *Encyclopédie* refer to the reprinted edition, *Encyclopédie ou Dictionnaire raisonné des sciences, des arts et des métiers* (Stuttgart: Bad Cannstatt, 1995). For *baguette divinatoire*, II:13; for *rabdomancie*, XIII:735–7. Also see François Ilharat de La Chambre, *Traité de la véritable religion*, 5 vols. (Paris: Hippolyte-Louis Guerin, 1737), II:469–85. Also see Kay S. Wilkins, "The Treatment of the Supernatural in the *Encyclopédie*," *Studies on Voltaire and the Eighteenth Century* 90 (1972): 1757–71.

31 For a summary of Bléton's early life see Louis Figuier, *Histoire du merveilleux*, II:104–5.

32 Scientists often considered the credulity of the general public as a chief reason for the success of charlatans: see Jean-Pierre Goubert, "The Art of Healing: Learned Medicine and Popular Medicine in the France of 1790," in *Medicine and Society in France*, eds. Robert Forster and Orest Ranum, trans. Elborg Forster and Patricia M. Ranum (Baltimore: Johns Hopkins University Press, 1980), 11. Nonscientific issues also fell under public scrutiny at this time. See Sarah Maza, *Private Lives and Public Affairs: The Causes Célèbres of Prerevolutionary France* (Berkeley: University of California Press, 1993). On the connections between science and the public sphere see Colin Jones, "The Great Chain of Buying: Medical Advertisement, the Bourgeois Public Sphere and the Origins of the French Revolution," *American Historical Review* 101 (1996): 13–40.

33 On medical electricity see John Heilbron, *Electricity in the Seventeenth and Eighteenth Centuries: A Study of Early Modern Physics* (Berkeley: University of California Press, 1979); Isaac Benguigui, "La théorie de l'électricité de Nollet et son application en médecine à travers sa correspondance inédite avec Jallabert," *Gesnerus* 38 (1981): 225–35; and Geoffrey Sutton, "Electric Medicine and Mesmerism," *Isis* 72 (1981): 375–92. On Mesmer see Darnton, *Mesmerism*; Charles Gillispie, *Science and Polity in France at the End of the Old Regime* (Princeton: Princeton University Press, 1980), 261–89; and Lindsay Wilson, *Women and Medicine in the French Enlightenment: The Debate over "Maladies des Femmes"* (Baltimore: Johns Hopkins University Press, 1993), chp. 5. On both topics see Brockliss and Jones, *The Medical World of Early Modern France*, 574–6 and 794–802.

34 Pierre Thouvenel, *Mémoire physique et médicinal, montrant des rapports évidens entre les phénomenes de la baguette divinatoire, du magnétisme et de l'électricité* (Paris: Didot, 1781), 7 (appeal to Galileo).

35 For "animal electricity" see *Nouvelles de la république des lettres et des arts*, 29 May 1782, 153–4. This is not necessarily related to Galvani's views on animal electricity.

36 Mesmerists also utilized the tactic of collecting affidavits and testimonials in order to support their claims and, like Bléton, Mesmer often took his appeals directly to the general public. Wilson, *Women and Medicine*, 113.

37 Journal des Savants, September 1781, 628. Guyton de Morveau, in *Journal de Nancy*, 3 (1780), 244–72 and 327–37; 4 (1781), 87–107 and 125–56. Gabriel-Antoine de Lorthe, "Lettre à M. Thouvenel," in *Mélange d'opuscules mathématiques*, 6 vols. (Paris: l'Auteur, Hôtel d'Orléans, rue Dauphine; J.P. Duplain, Libraire, Cour du Commerce, 1782–1785), 7; J.P. Battelier, in *Nouvelles de la république des lettres et des arts*, 12 June 1782, 169. Battelier published a series of letters in this journal; see also ibid., 19 June 1782, 177–9; ibid., 26 June 1782, 186–88; and ibid., 3 July 1782, 193–6.

38 Sigaud de la Fond, *Dictionnaire des merveilles de la nature*, 2 vols. (Paris: Chardon, 1781), I:75, 76.

39 Bachaumont, *Mémoires secrets pour servir à l'histoire de la République des Lettres en France depuis 1762 jusqu'à nos jours*, 36 vols. (London: John Adamson, 1784–1789), 20 (1782):248–50.

40 *Journal des Savants*, August 1782, 558–64. Henri Decremps, *La magie blanche dévoilée*, (Paris: Langlois, 1784–1785), 72–3. Other authors copied Decremps' explanation. See, for

example, Jacques Lacombe, *Dictionnaire encyclopédique des amusemens des sciences mathématiques et physiques* (Paris: Panckoucke, 1792), 267–9.

41 On enthusiasm, *Nouvelles de la république des lettres et des arts*, 12 June 1782, 170; on public opinion, ibid., 3 July 1782, 193.

42 *Journal de Paris*, 13 May 1782, 531. See also *Journal encyclopédique*, July 1782, 142–4. Guillaumot had extensive experience with underground objects himself. At about the same time that he put Bléton to the test, he also supervised the construction of the Catacombs.

43 *Journal de Paris*, 21 May 1782, 564; ibid., 26 May 1782, 583–4.

44 The full account, published in the *Journal de Physique*, came complete with a diagram of the gardens with Bléton's route mapped out point by point; *Journal de Physique*, 20 (1782): 58–72 and plate. *Journal de Paris*, 16 June 1782, 675–7.

45 For Thouvenel's reaction to the experiments see *Journal de Paris*, 2 June 1782, 612–13; *Monthly Review*, 67 (1782): 553–6; and *Journal des Savants*, 20 (1782), 70. For the new results see *Journal de Paris*, 26 June 1782, 719–26. This supplement included the results of some more experiments as well as letters written by Parisian supporters of Bléton including a minor academician (Louis Cotte) and a royal official (Le Roi). Guillotin was also on the commission that condemned animal magnetism but he came down in favor of dowsing.

46 For a description of this experiment see *Journal des Savants*, August 1782, 561; and Barrett and Besterman, *The Divining Rod*, 45. Barrett and Besterman point out that Charles' experiment was flawed, albeit in a way not understood at the time. The insulated stool would not have prevented any underground electrical influence from affecting Bléton; it would have prevented electric conduction but not electric induction.

47 For the association by mesmerists with dowsing, see Joseph-Jacques Gardane, *Eclaircissemens sur le magnétisme animal* (London, 1784), 6–7; Jean Jacques Paulet, *Réponse à l'auteur des doutes d'un provincial, proposés à MM. les Médecins-Commissaires, chargés par le Roi de l'examen du magnétisme animal* (London, 1785), 19; *Journal de Paris*, 22 February 1784, 242–3; Joseph Philippe François Deleuze, *Histoire critique du magnétisme animal*, 2nd edn, 2 vols. (Paris: Belin-Leprieur, 1819), II:277–80; and the anonymous pamphlet, *Le Cri de la nature, ou le Magnétisme du jour* (London; Paris: Chez les Marchands de Nouveautés, 1784), 15–16. For a combined attack on mesmerism and dowsing see *Journal de Nancy*, 14 (1784), 112–29.

48 On the anti-dowsing side see for example *Journal encyclopédique*, August 1782, 145–52; ibid., September 1782, 491–501; ibid., March 1784, 309–20; and ibid., July 1784, 307–17. On the pro-dowsing side see for example *Journal de Paris*, 6 September 1782; and the lengthy excerpts found in ibid., 16 June 1785 to 20 June 1785. On ballooning and dowsing see *Journal de Paris*, 4 January 1784, 14.

49 See the articles on the *baguette divinatoire* and *hydroscope* in Gaspard Monge et al., *Dictionnaire de physique*, 4 vols. (Paris: Hôtel de Thou, 1793–1822), I:2–9, III:500. Also, see the bibliography in Barrett and Besterman, *The Divining Rod*.

50 *Journal des Savants*, 25 (1784): 314–15. Compare this to the review in the *Gazette de Santé* no. 47 (1784): 185–7. On the coalfield see *Mémoires secrets*, 26 (1784): 91–2.

51 Journal de Paris, 17 June 1785, 695.

52 *Journal de Nancy*, 6 (1781), 91.

53 Barbara Benedict, *Curiosity: A Cultural History of Early Modern Inquiry* (Chicago: University of Chicago Press, 2001), 159.

6

Balloons and mass science

In 1783, a "sublime invention" shocked France and the rest of the world. The Montgolfier brothers invented the hot-air balloon. Within a short period animals, and then humans, went aloft in hot-air and hydrogen balloons. This instigated an enormous craze for balloons and marked the first time that humans had defied gravity and left the earth to roam through the heavens. Balloons became an emblem of the age of Enlightenment and quickly came to occupy the minds and thoughts of scientists and amateurs alike. Francis Hopkinson claimed, in a letter to Thomas Jefferson, that, in the newly created United States, for every "one person that will mention [Congress] a hundred will talk of an Air baloon [sic]." Launches were highly public affairs attended by massive crowds who often paid for the privilege of watching balloons take off. This is an early instance of mass culture with socially mixed groups all participating in a single scientific event both as patrons and witnesses.[1]

Contemporaries touted balloons as one of the most important inventions of the eighteenth century. One anonymous author suggested that ballooning held a place of honor alongside inoculation while another, somewhat more problematically, claimed that the discoveries of ballooning and animal magnetism best illustrated the successes of the eighteenth century. Another individual noted that while the century of Louis XIV "was destined to perfect the arts and letters . . . the great discoveries in the high sciences were reserved for the century of Louis XVI." This author pointed to the work of Benjamin Franklin, Jean-Antoine Nollet, Comus, Mesmer, Cook, and, of course, Montgolfier, and specifically cited electricity, animal magnetism, the discovery of how to calculate longitude, and the invention of balloons. Significantly, this list concentrates largely on sciences that provided a great deal of spectacle and on those many savants who specialized in public performances. In any case, as a reporter for the *Courier de l'Europe* suggested, balloons clearly occupied the "physicists, mechanics, and all the savants" of Europe; even the people, the writer suggested, sensed that ballooning was "all important."[2]

This chapter focuses on the manner in which balloons underwent popularization and commodification.[3] In the 1780s, ballooning "fixed the attention of all the savants, and became the unique object of conversation in all assemblies." The French discussed ballooning in the city and at court and in "all our circles, at all our suppers, in the toilettes of our young women, as in our academic lycées."[4] This scientific discovery spread rapidly from the provinces to Paris and on to the rest of Europe.[5] I argue that public participation in balloon launches, most importantly through subscriptions but also through the purchase of balloon-themed material goods, transformed this scientific endeavor. Aeronauts exerted considerable effort in gaining popular support and approval while simultaneously focusing less on the scientific potential of balloons. The practice of funding balloons through subscriptions gave the general population a much greater influence over the practice of this science than had ever been the case. Thus, after a brief examination of its origins, I turn to an analysis of the spectacular and commercial side of ballooning. The last part of this chapter examines ballooning as an early example of mass culture and explores the behavior of the crowds and attempts to police them by the state.

Balloons affected all aspects of culture. Balloon launches were viewed as an enormous spectacle, one that inspired poets and playwrights to put in writing, for better or worse, their thoughts and emotions. Merchants, however, also got into the act and ballooning quickly became a commercial event sponsored through subscriptions and available for the everyday consumer through the sale of small balloons and other objects. As with divining rods, a debate arose over the utility of ballooning. Proponents of ballooning, accepted as scientific by just about everybody, had trouble justifying their interest in an endeavor that failed to have an easily identifiable use.

The spread of ballooning

According to Louis-Sébastien Mercier, 1783 "was a year of marvels": in addition to balloons he noted that Comus opened his Hospice Médico-Electrique, Mesmer returned to Paris, and Spallanazi published his work on digestion.[6] Ballooning, however, clearly dominated the headlines of popular science. The public story of ballooning begins on 4 June 1783 when the Montgolfier brothers, Joseph and Etienne, performed a successful balloon launch from the Place de Cordeliers in Annonay, France. The balloon stayed aloft for approximately ten minutes, attained a height of around three thousand feet, and traveled about a mile and a half before crashing onto a stone fence in a vineyard and catching fire. This experiment took place before a number of spectators, most important of whom were the members of the Etats particuliers, the diocesan assembly of Vivarais region who were in Annonay for their annual meeting. This group, at the request of the Montgolfiers, drafted a formal letter concerning the experiment that they sent to the Académie Royale des Sciences. This letter helped Etienne and Joseph establish the priority of their balloon launch and ensured the accolades and notoriety they felt they deserved. Fame came quickly as news of this event rapidly reached Paris where it created a sensation. One commentator claimed that this discovery was comparable to those of Christopher Columbus and Nicolas Copernicus. Another author related it to even more disparate events such as the eradication of prejudice and superstition, the elimination of slavery in Europe, the creation of the United States of America, and the suppression of the Jesuits.[7]

The idea of conquering gravity, of course, was not original to the Montgolfiers. Commentators on balloons were quick to recall the work of the seventeenth-century author Savinien Cyrano de Bergerac, for example, who had written a futuristic pair of novels, *Voyages to the Moon and the Sun*, in which the protagonist had attached jars of lighter-than-air gases to his body and leapt higher and higher until making it all the way to the heavens. As Awen A.M.Coley has shown, the eighteenth century teemed with tales of flight, based on classical mythology and bolstered by scientific discoveries. In addition, just two years before the Montgolfier brothers made history, Jean-Pierre Blanchard had proposed, in the pages of various newspapers, a "flying ship" [*vaisseau volant*] but had, to his dismay, never been able to build a successful model.[8]

While the Montgolfiers had used hot-air to inflate their balloon, the first balloon launch in Paris used hydrogen. The well-known scientific popularizer Jacques-Alexandre-César Charles and two renowned makers of scientific instruments, the brothers Anne-Jean and Nicolas-Louis Robert, oversaw this event. Hydrogen allowed for a smaller balloon, since it has more lifting power than hot-air, but also resulted in new problems, not the least of which was how to manufacture the necessary nine hundred cubic feet of the gas.[9] After much trial and error Charles and the Roberts inflated and launched their balloon on 27 August 1783.[10] This balloon, which at twelve feet in diameter was much smaller than its predecessor, took off from the Champ de Mars and quickly drifted out of sight. It landed forty-five minutes later about 20 kilometers outside Paris, where the peasants of the village of Gonesse attacked it with pitchforks out of fear for their lives.

Following the initial success of these launches, balloonists quickly pushed for a more elite audience and more daring demonstrations. In particular, the Montgolfier brothers wanted to perform a launch for the court in order to obtain royal patronage and, they hoped, the legitimacy that could only be granted through a royal audience; plus they wanted to see what would happen to living creatures sent up in balloons. Both these goals were met on 19 September 1783 when, at Versailles, Etienne Montgolfier launched a balloon before Louis XVI that included three passengers – a rooster, a duck, and a sheep. All three animals survived their adventure; Marie-Antoinette subsequently honored the sheep, which she dubbed Montauciel, with a place in the royal menagerie. It now seemed logical to inquire whether humans could ascend in hot-air balloons, although some people thought this possibility too dangerous. In one newspaper, for example, the authors noted that the inability to steer balloons made them too uncertain and, therefore, too hazardous to the lives of the men who might go up in them. In a telling bit of social commentary, this same journal later suggested that the first person to go up in a balloon should be an "intelligent worker."[11]

In spite of opposition, the experiment took place and the first person to go up in a hot-air balloon was the daring Jean-Francois Pilâtre de Rozier, scientific popularizer and head of the Musée de Monsieur. He ascended, along with the Marquis d'Arlandes, on 21 November 1783 from the château de la Muette, home of the official court of the royal children. While Pilâtre de Rozier has long been acknowledged as the first man to ascend in a hot-air balloon, Charles Gillispie argues, based on a close reading of the Montgolfier family correspondence, that Etienne Montgolfier most likely had already gone up in a tethered balloon several times during trials prior to Pilâtre de Rozier's more public launch. Significantly, however, the presence of members of the general public as witnesses meant that Pilâtre de Rozier could claim all the honors. In any case, this feat was followed on 1 December 1783 by the first manned flight of a hydrogen balloon by Charles and Nicolas Robert from the Jardin des Tuileries with approximately 400,000 people, or half the population of the city of Paris, in attendance.[12]

Having arrived in Paris from the provinces, ballooning now quickly returned to the countryside, as well as to other countries, where balloon launches proliferated. Experiments in cities such as Lyon, Avignon, Dijon, Marseilles, Rouen, Aix-en-Provence, and Bordeaux took place over the next year as savants, nobles, and entrepreneurs all began to dabble in this new science. Many of these flights stayed aloft for only a short time; but some achieved outstanding success. The flight arranged and manned by three savants in Bordeaux – Darbelet, Desgranges, and Cahlifour – traveled approximately seventy kilometers in under two hours on 26 July 1784. Jean-Pierre Blanchard, along with a professor of anatomy named Dr. John Sheldon, flew almost seventy-three miles on his fourth flight that went from Chelsea, just outside London, to the town of Rumsey. On 7 January 1785 Blanchard, along with Jeffries, made the first trip in a hot-air balloon across the English Channel from Dover to Calais, a flight that took them just under three hours. On 15 June

1785, an attempt by Pilâtre de Rozier and the artist Romains to go the other direction ended in disaster when their balloon caught fire. Thus, Pilâtre de Rozier earned yet another place in history as the first man since Icarus to die as a result of falling from the sky. While this disaster put a damper on things for a while, balloons remained a popular public experiment well into the nineteenth century.

Balloons also quickly attracted attention throughout the rest of Europe and even further abroad. The first balloon launch in the Americas, on a plantation outside the city of Cap François in Saint Domingue, occurred in 1784 only seven months after the Montgolfier brothers made their first public flight. Several more flights on Saint Domingue took place over the following months much to the delight of many of the inhabitants. One commentator noted that this "novelty pleases women" in particular. Balloons were also launched in the Spanish province of Oaxaca, now part of Mexico, in 1785. These launches predated the first successful manned flight in the United States, accomplished in Philadelphia in 1793 by Jean-Pierre Blanchard. England saw an enormous number of balloon flights, some performed by foreigners like the Italians Vincenze Lunardi and Count Zambeccari and the ubiquitous Blanchard. Balloons also quickly reached Ireland where a number of experiments were conducted, several by Richard Crosbie and even more by a Scottish scientific lecturer who frequently traveled in Ireland, James Dinwiddie.[13]

Public support and the commercialization of balloons

Almost immediately after the invention of balloons, people proclaimed their scientific importance while also viewing them as a social and financial opportunity. In other words, people could acquire cultural capital from knowledge of balloons, from possession of aeronautic goods, or even from simply attending balloon launches. With strong public interest in this new invention, it is not surprising that the Montgolfier brothers initially sought to support their efforts through public subscriptions. But this provided neither the honor nor satisfaction they desired. Thus, after a brief foray into public support, they turned their attention to the crown. They wanted recognition and accolades from the state and its scientific institutions, not from the general public. Most other people involved in ballooning, however, sought public support using, in particular, the sale of tickets to launches.[14]

A few subscriptions targeted an elite audience. The expenses incurred for a balloon experiment could be quite high; thus, subscriptions might fund even moderately priced launches, such as the one in Rodez in August of 1784 which cost a mere 1,800 livres.[15] The subscription set up to finance the Montgolfier brothers' 1784 launch in Lyon, for example, cost twelve livres per person; the goal was to get 360 people to subscribe. Ultimately, they sold 283 subscriptions to 174 people for a total of 3,396 livres. The Prince de Ligne and the duc d'Ursel together purchased one hundred tickets while sixty-five members of the Cercle des Terreaux anonymously purchased tickets for themselves. The rest of the subscribers bought individual tickets. Women, mostly those married to other subscribers, bought eleven of these. Only one woman, a Mlle Salade, was listed as unmarried; she, however, became the target of satirists for purchasing a ticket. Critics described her as vain, and ridiculed her for rising above her station, since she was the proprietor of a small restaurant [*marchande gargotière*]. Where indicated, the status of subscribers tended to be fairly elite with counts, barons, and other nobles represented along with the local intendant, lieutenant of police, tax farmers, and other office holders. Savants also participated, with ten academicians listed. Only around ten percent of the subscribers appeared to be of bourgeois or merchant status. Most of these people lived in the area although a mysterious M. Gretthet, an "English gentleman," also subscribed.[16]

While this flight ultimately succeeded, it did suffer a series of problems. Uncooperative weather and problems during the inflation of the balloon resulted in a delay of several days, spent frantically patching holes. When the balloon was finally ready, and the weather nominally clear, the organizers then faced the problem of who would go up in the balloon. The list of individuals who wanted to go seemed to get longer and longer with each passing day, a reality noted even in the newspapers. Several aristocratic subscribers felt that they had the right to go and so Prince Charles, the marquis de Dampierre, the comte de Laurencin, and the chevalier de Laporte d'Anglefort all leapt aboard. Joseph Montgolfier, who was technically in charge of the launch, wanted to go himself and he had also promised Pilâtre de Rozier a place. He was worried, however, that the balloon could not carry so many people. The noblemen, who were all armed, drew their pistols and threatened to kill anyone who tried to make them leave the gondola. While the local intendant, Flesselles, and others urged them to listen to reason they instead insisted that the cords holding the balloon be untied. At this point, Pilâtre de Rozier and Joseph Montgolfier both got into the balloon. Just as the balloon was about to take off, a seventh, uninvited, aeronaut named Fontaine leapt on board. The balloon, now much too heavy, rose slowly before the spectators and ascended to about 2,400 feet, but the entire flight lasted only about fifteen minutes.[17]

Several balloonists opened equally expensive subscriptions aimed at the French elite. When Pilâtre de Rozier opened a subscription to fund his first flight in a Montgolfier-designed balloon, he received 1,200 livres from his noble supporters and opened a subscription to fund the rest. The subscriptions cost twenty-four livres and came with four tickets for viewing the launch.[18] Other balloonists adopted this model, such as in a 1784 launch in Dijon. For twenty-four livres the subscriber received four tickets to the launch as well as admission to a second launch to take place some time later. The balloonists invited individuals who purchased three subscriptions to attend all the preliminary work in the construction and inflation of the balloon and printed their names along with a description of the flight after the experiments were over.[19] Different subscriptions came with other perquisites. The *Courrier d'Avignon* announced one in which the price of the ticket allowed the subscriber to visit the workshop where the balloon was to be constructed. They could also twice visit the model balloon being displayed as part of this project. To advertise their subscription, Charles and the Robert brothers put their balloon on display several days in advance, along with the rest of the equipment. Their enormously successful subscription may have earned them as much as fifty thousand livres. While this claim may be overstated, Charles and the Robert brothers clearly earned more than just respect for their efforts.[20]

While a few balloonists targeted the wealthy, others set their subscriptions at a lower rate, usually three livres or less. In other words, most aeronauts tried to garner patronage from a large number of people each paying a small amount. This is an essential shift in ballooning, and popular science in general. By targeting a broader audience, these balloonists placed their status, economic livelihood, and future in the hands of the public in ways that their elite counterparts did not wish to do. Charles Gillispie suggests that some balloonists avoided public support for their work to prevent the public from having too much input into when and where launches would take place. From his point of view, the public's interest in balloons was regrettable since it prevented aeronauts from accomplishing anything of real scientific value.[21] However, there were only a few balloonists who could afford to eschew public support altogether.

Many balloonists depended on support from a larger public. The Robert brothers performed an experiment in September of 1784, along with their brother-in-law Collin-Hulin, for which they sold advance tickets at three livres each. Monsieur Nicolas, a professor of chemistry in Nancy, opened a subscription at three livres for his balloon launch in that city in late 1783. Monsieur Brun opened a subscription in 1787 hoping to raise two thousand livres. He offered two options: individuals could pay three livres for the best seats or thirty sous for lesser seats. Alban and Vallet announced, in a short précis, their 1785 launch in the town of Javel. Anyone who bought a copy of this announcement, at three livres each, could view the launch and gained one-time admission to see the balloon and its accompanying apparatus in advance. Alban and Vallet also offered a more extended course on ballooning. This course, priced at five gold Louis, was much more costly, but it included many amenities. Subscribers, limited to four hundred, could stop by ten times, between 6:00 a.m. and 1:00 p.m. and again from 5:00 p.m. to 9:00 p.m., to see the construction of the balloon during the month prior to the launch. They could also climb into the gondola, receive a complete explanation of how balloons worked, and have seats to watch the launch.[22] Tétu de Brissy, another man who used subscriptions to help launch balloons, offered several different experiments between 1783 and 1786. Initially, he sent off unmanned balloons. Eventually, however, he himself went up in a balloon. He had hoped to set a record and stay aloft for twenty-four hours but only stayed up around five hours.[23]

Individuals could purchase tickets for balloon launches from a variety of places. Balloonists, of course, sold them in their workshops and homes. Subscriptions were typically set up in public places in order to spread the word as widely as possible about the chance to invest money in a balloon. The subscription for Charles and the Robert brothers' hydrogen balloon could be found in the Café du Caveau in the Palais Royal, a location which ensured that many people, both during the day and at night, would learn of it. The *Courrier d'Avignon* claimed that among those who subscribed were "princes, ministers, men of letters, [and] citizens of all social levels [*états*]." Tickets for Brun's balloon in Nancy could be purchased at the office of a printer while tickets for Nicolas's balloon launch in the same city could be acquired at a local apothecary.[24] Subscribers bought their tickets in locations that served as centers of exchange and commerce. Tickets for a subscription announced during the French Revolution were available for purchase at the shop of the widow Tourilon, a tapestry merchant, at the Lycée de Palais Royal, and at the Panthéon.[25]

It is unsurprising, perhaps, that some balloonists were accused of being overly fixated on making money. There was, in fact, at least one aeronaut – namely, Jean-Pierre Blanchard – who made a career out of his predilection for balloons and became the biggest of the commercial balloonists. Blanchard eked out a living as a professional aeronaut and earned for himself the nickname "Don Quixote of the English Channel" for his daring deeds. Between March 1784 and his death early in 1808 Blanchard ascended approximately sixty times in balloons around Europe and in North America. Although impoverished at the time of his death, he had survived largely on his ballooning expeditions for nearly twenty-five years, thus making him one of the first long-term aeronautic entrepreneurs. He did this so well, in fact, that one contemporary suggested that Blanchard was more interested in profit than in anything else, a characterization that may well be true.[26]

Blanchard's father was a turner and a cabinetmaker, a profession that Jean-Pierre initially followed. He also had a talent for machines and mechanical devices and, at the age of sixteen, created a mechanical carriage. He left home in 1781 and set out to make his fortune in Paris where he worked on his flying ship. He garnered enough attention for his invention to gain the patronage of the king's younger brothers, the comte d'Artois and the duc de Chartres (the future duc d'Orléans). His failures, on the other hand, warranted an attack from the academician Joseph Jérôme Lalande who in 1782 called his invention inane. Blanchard became something of a laughing stock and the subject of a theatrical farce, *Cassandre Mécanicien, ou le bateau volant*.[27] By the time the Montgolfier brothers tested their first balloons, Blanchard was almost thirty years old. He had already caused a furor in Paris by claiming to have created a "flying ship," a sort of precursor to the balloon modeled after sailing ships. He raised a subscription to help fund his experiments but after numerous attempts, rain delays, and other problems he never got it off the ground. After the success of the Montgolfier brothers Blanchard tried to claim that he really deserved precedence for the invention of balloons, a claim no one took seriously. He later tried to take credit for the invention of the parachute as well.

A dour and humorless man (one of his nicknames was "Pisse tout dret [droit]"), Blanchard nonetheless capitalized on the invention of ballooning and became one of the most recognizable figures in Europe. He launched his first balloon in Paris in March 1784 and followed this up with two launches in Rouen in May and July. He then took off for Great Britain where he performed in London. As noted earlier, on 7 January 1785 Blanchard and an American artist named John Jeffries traveled from Dover to Calais. This feat earned Blanchard an enormous reputation and sealed his fate. He spent the rest of his life planning for his next launch. He did this in places as varied as Rotterdam, Liège, Brussels, Metz, Strasbourg, Nancy, Leipzig, Warsaw, Vienna, Prague, and Hanover. He was the first person to ascend in a balloon in North America when he made his flight on 9 January 1793 in Philadelphia in the presence of a large crowd that included George Washington. In addition to the United States, Blanchard is credited with the first flight in a German-speaking area (3 October 1785 in Frankfurt am Main), in Belgium (20 November 1785 in Gand), in Switzerland (5 May 1788 in Bâle), and Poland (30 May 1789 in Warsaw). In 1785 Blanchard tried to establish the Académie Aérostatique in Paris. This may have been an attempt to both rival Pilâtre de Rozier's Musée de Monsieur and to provide himself with better credentials and patronage to help counter his critics. In any case, nothing seems to have come of this effort.[28]

Blanchard often flew solo but he would occasionally take passengers with him. In London, Miss Simonet ascended on 3 May 1785 and her younger sister went five days later. Noblemen also rode with him, oftentimes purchasing multiple tickets for the privilege. This was the case when the chevalier l'Epinard rode with Blanchard in a seven-hour flight from Lille to Servon. In 1799 the ten-year-old Mlle Maison traveled with Blanchard and, that same year, Lalande, who had criticized Blanchard's flying ship seventeen years earlier, took off with him from the Jardin du Tivoli. Dogs made the ascension several times; they also performed valuable experiments, although it seems likely that their participation took place unwillingly. On 3 June 1785, for example, Blanchard threw a dog out of the gondola attached to nothing but a parachute. The dog lived to tell the tale and Blanchard repeated this experiment several times. Later, Jacques Garnerin, who had also experimented with parachuting dogs, became the first human to successfully use a parachute.[29]

Not all the passengers in balloons participated with permission. For his first flight at the Champ de Mars in 1784, Blanchard had planned to take off with a Benedictine monk named Dom Pech. Pech's superiors, however, had forbidden him from participating in the launch although he arrived to do so anyway. Just before the launch, a young cadet named Dupont de Chambon arrived on the scene and demanded that Blanchard allow him to ascend as well. When Blanchard refused, the disgruntled youth drew his sword, punctured the balloon, and stabbed Blanchard in the hand. In the ensuing ruckus, he broke many of the scientific instruments as well as the mechanism Blanchard had designed to try to steer the balloon. Blanchard, despite his injury, went through with the flight although the damage to the balloon meant that as fate would have it Dom Pech could not.[30]

In addition to taking passengers along with him on his flights, Blanchard also made ballooning a family affair. In 1774 he had married Victoire Lebrun, with whom he had four children. However, he abandoned Victoire to travel about Europe for his ballooning career and she died in poverty. He married his second wife, Marie-Madeleine-Sophie Armand, in 1797 when he was forty-four and she just nineteen; she subsequently followed her husband into the ballooning business. When Blanchard died in 1808 he was not only poor but in debt. Marie-Madeleine-Sophie vowed to pay his creditors and continue as a professional balloonist, goals she successfully met for the next eleven years. She became a regular fixture in Paris with her extraordinary balloon launches. Most famously, she took off twice each week from the Jardin du Tivoli, a public pleasure garden in Paris formed from the estate of one of the many aristocrats who had fled during the French Revolution. Her popularity continued after the fall of Napoleon and she became the official "aeronaute of the Restoration." On 6 July 1819, she tried to offer a special treat at a festival held at Tivoli and attached a number of fireworks to the gondola of her balloon, a trick she had performed previously but never on so grand a scale. Her balloon caught fire and crashed into a house at 15, rue de Provence. Marie-Madeleine died after being thrown from the gondola onto the roof of the house and from there to the street below.[31]

Many individuals tried to capitalize on the mania for balloons. Like Blanchard, a number of other balloonists used subscriptions to garner profits. Campmas, for example, who called himself an engineer, planned several launches in 1784. He sold tickets at a variety of locations around Paris including a restaurant and several cafés. The price ranged from three to twelve livres. The latter price enabled the bearer to attend all the planned launches. Those who wanted to attend one launch paid the lesser fee. Another entrepreneur, Enslen, who hailed from Strasbourg, traveled around Europe and launched unmanned balloons. A mechanic and a physicist who called himself the "Ariste de Strasbourg," Enslen performed this feat in Lille and Metz as well as in Paris. He created balloons shaped into various mythological forms such as a nymph and, predictably, Pegasus. While in Metz, he set up a display of his "aerial figures" admired, he claimed, by the best physicists in Paris. Access to these figures cost twelve sous per person, with children getting in for half price.[32]

In Paris, Enslen teamed up with the well-known experts in fireworks and operators of a pleasure-garden on the rue St-Lazare, the Ruggieris. The Ruggieri brothers came from a long-line of artisans specializing in fireworks. Beginning in 1766, the Ruggieris ran a garden where a small fee gave individuals access to music, food, and fireworks displays. After 1783 they offered balloon launches as well. The price of admission for these special events varied depending on proximity to the action. It cost three livres to sit in the loge or gallery while a place in the garden cost only thirty sous. Advance tickets could be purchased at the Café des Arts on the rue de Tournon and in a boutique at the Palais Royal indicating, along with the price, that these entertainments were designed for the relatively well-to-do. One contemporary claimed that "all Paris" had been to see the balloon figures where they were housed in the Palais Royal; the price to view them in advance was a mere twelve sous.[33] Aside from featuring Enslen's launches, the Ruggieris also attempted a few balloon experiments on their own. In 1789 they launched a balloon and demonstrated a parachute based on the model provided by Blanchard. At one point they announced a set of launches, the proceeds of which were to benefit the poor, although the hospital had a difficult time getting any money out of them.[34]

Within a short span of time, balloons became enormously fashionable and dominated late eighteenth-century scientific culture. In addition to the vast number of people who went to see balloons and balloon launches, even more people purchased ballooning merchandise. Enterprising merchants offered novelty miniature balloons for sale in their boutiques, designed for children, as well as larger versions. These sold for as much as six livres and as little as forty sous, depending on the size. One woman apparently made a balloon skirt out of taffeta, although it is unclear if this commodity was for sale. Another merchant designed balloon hats and sold ribbons "à la Montgolfier." Some women styled their hair after balloons and men could purchase shoe buckles in the shape of a balloon. As early as September 1783, Marie-Thérèse-Charlotte, the young daughter of Louis XVI and Marie-Antoinette known usually as Madame Royale, arrived at a balloon launch wearing a "pretty little hat 'à la Montgolfière.'" Balloons appeared on plates, clocks, chamber pots, furniture, and all sorts of other household goods. There was even a balloon game and a balloon liquor (crême aérostatique). Restaurants cashed in on the craze with one offering a filet à la Montgolfier. A wax museum added a figure of Jean-Pierre Blanchard displayed alongside Benjamin Franklin, the comte de Buffon, Jean-Jacques Rousseau, and Franz Anton Mesmer. Choosing Blanchard for this honor, rather than the Montgolfier brothers, Pilâtre de Rozier, or Charles reveals the popularity enjoyed by Blanchard in spite of his position as a merchant of ballooning rather than as a ballooning savant.[35]

Mass science and public spectacle

While aeronauts and merchants commercialized balloons in new ways, they also integrated them into the existing superstructure of popularized science. At the same time, the overall size and scope of popular science underwent a major transformation as ballooning drew huge crowds and enormous public interest. Disseminators, at least those who wanted to keep up with the times, quickly incorporated ballooning into their public lectures or wrote pamphlets and books on the subject. Bienvenu, who manufactured and sold scientific equipment in addition to offering short courses on how to use the instruments, sold an assortment of balloons made from tafetta or *baudruche* (a film/fabric made from sheep or cattle intestines) that individuals could purchase at his shop or, in the case of people living in the provinces, have sent to them. He also offered a series of six lectures on gases, combined with lectures on electricity. Bienvenu offered this course free of charge and, purportedly, designed it so that it would optimally demonstrate the machines he sold in his shop. He announced this course for the first time in early 1784, less than six months after the Montgolfier brothers made their first flight; Bienvenu quickly adapted his efforts at popularization to address the commercial possibilities of this new trend.[36] Monsieur Lebrun, claiming to be the chief of balloons for the royal children, sold equipment designed for people who wanted to construct their own balloons. He marketed these goods out of his shop on the rue de Richelieu but was also willing, he said, to ship them to the provinces.[37]

Other popularizers also entered this new market. For his course that opened on 4 December 1783, Antoine Deparcieux noted that he was sure to offer many lectures dedicated to experiments with balloons. Deparcieux appealed to a broader social group by offering two versions of this course; the first met at 11:00 a.m. and the second met at 6:00 p.m.[38] This allowed a more diverse group of people to learn about balloons. Rouland began mentioning balloons in his advertisements as early as September 1783. He already offered a course on air and electricity and easily integrated balloons into his lectures.[39] Popularizers who themselves went up in balloons, such as Charles and Pilâtre de Rozier, did not need to advertise any more than usual since their deeds appeared so frequently in the newspapers.

The science of ballooning provided disseminators with the quintessential object of entertaining and instructive popular science.[40] The launch itself could easily be equated with a theatrical spectacle. It was choreographed; it occurred at a specific, predetermined time; there was an audience, many of whom had purchased tickets priced according to the quality of the seats; there were performers with props; and, sometimes, problems erupted, much to the joy or irritation of the audience. During one experiment in Nantes in 1784, "warlike music," designed to mimic the journey of the balloon, played throughout the period of the launch. When Charles took off from the Champ de Mars in 1783 the event was marked with cannon fire to announce his departure.[41] The launch itself inspired conflicting emotions among those present. Charles's launch from the Champ de Mars was accompanied by "repeated applause" that, according to the author of an article for the *Journal de Paris*, provided "new proofs of the interest of the public" in the sciences and the arts. Hours before the balloon launch at Versailles, the road to Louis XVI's palace was filled with carriages. People packed the courtyard and the windows overlooking the launch site.[42]

An experiment in Lyon, however, proved more ambiguous. The emotions of the spectators ran wild and they were filled "with joy and with fear" while they emitted cries and wrung their hands.[43] At a different launch some spectators fell to their knees in prayer, while others stood up, raised their hands and cheered. Some spectators even fainted. An early launch from the Tuileries occurred amid massive silence and admiration; the Baroness d'Oberkirch wrote that the crowd who witnessed a launch with her was "stupefied."[44] Count Louis Philippe de Ségur later wrote that although the spectacle of ballooning had become, by the early nineteenth century, "a vulgar ornament of our fêtes," his first sight of a balloon produced a "vivid and profound effect" on him.[45] So great was the demand for balloons at festivals, wrote another commentator, that "the workers of Paris could not furnish the demand for the provinces."[46] Spectators could be found everywhere including on the streets, in the gardens, peering out of windows, and on rooftops. Notre Dame Cathedral became, temporarily, a viewing platform for the balloon launched from the Château de la Muette in 1783. The observers were treated to a minor eclipse when the balloon traveled between them and the sun. Proprieties were often ignored in the effort to get a good look. Guardians of a church near the fair at Saint-Laurent charged admission to the cemetery where people could watch while seated alongside numerous insufficiently buried corpses. When the balloon failed to launch, these spectators adamantly demanded reimbursement.[47]

The extreme responses to the spectacle of ballooning led some commentators to describe it as a "mania" dominated by the enthusiasm of the spectators. They found it particularly scandalous that some women were so captivated by balloons that they stayed outside in the rain all day to guarantee their place at the launch. Those who could afford it, like the English traveler Anna Cradock, sent their servants ahead to save them a place.[48] The large number of people who came to see the balloon launches necessitated arrangements like these. These events were one of the first examples, albeit somewhat short-lived, of a truly mass culture thanks to the huge numbers of spectators drawn to launches and to the infiltration of balloons into so many different aspects of material culture. These occasions drew, at times, thousands of people. The largest, Charles's ascent from the Tuileries in 1783, drew an estimated four hundred thousand people, approximately half the population of Paris at that time. An observer noted that "both sides of the Seine were covered with a multitude of spectators" witnessing the launch. A hundred thousand people reportedly attended the first balloon ascension at Versailles. At other times and in other places crowds of ten thousand or more were not unusual.[49]

Some individuals concentrated on the reactions to balloons rather than to the launches. Several Japanese fishermen viewed a balloon launch with relative nonchalance. When questioned about their reaction they apparently replied that it was "nothing but magic and in Japan we have practitioners of magic in abundance."[50] Some observers suggested that peasants were reacting superstitiously to this invention. Charles described the peasants following his balloon "as children who followed the butterflies in the field." More typically, people described the peasants as reactionary. In one case, they apparently threw stones at the balloon.[51] Nicolas, a chemistry professor in Nancy, ridiculed the peasants mercilessly. He launched an unmanned balloon and trailed after it on the ground. When he asked the peasants if they had seen it they responded that they did not understand what he meant by balloon, but that they had seen a large bird. In writing about the incident he emphasized their lack of understanding and mimicked the local patois in his report, spelling the peasants' response to his question phonetically and then translating it for his readers.[52] Most famously, the peasants of the village of Gonesse were depicted in contemporary prints with pitchforks and other tools attacking a balloon that had the misfortune to land near them.[53]

Balloon launches, and the "immense crowd of spectators" who came to watch, became a part of the culture of late eighteenth-century France. Not everyone, however, thought that large crowds of people were a positive result of the popularization of aeronautics. Benjamin Franklin suggested that balloonists should take care; if they drew a large crowd but then failed to launch a balloon, dire consequences might follow. He then related a story from Bordeaux in which a savant failed to launch a balloon, after collecting money from subscribers. As a result an angry mob razed his house.[54] Mobs frequented more than just the sites of launches. A crowd surrounded Charles's home on the Place des Victoires after his first balloon demonstration. On another occasion, balloons, or more precisely the memory of a balloon launch, served to quiet a mob. During the French Revolution rioters stormed the Louvre and found themselves in the Galerie d'Apollon where Charles lived and tended to his scientific equipment collection after he had donated it to the state. As they were about to ransack the room, someone saw the gondola Charles had used in his ascent. Instead of looting, the crowd quieted down and asked Charles to recount his adventures in the balloon.[55]

Not all crowds could be quieted so easily. Such was the case with the infamous attempted balloon launch undertaken by the Abbé Laurent-Antoine Miollan and his partner Janinet in 1784. In the previous year, Miollan had begun a series of public lecture courses in experimental physics. Janinet, for his part, described himself as an "artisan-physicist." [56] They announced their balloon launch in the periodical press and, as usual, established a subscription, with tickets ranging from three to six livres. Problems arose, however, when they tried to inflate their balloon. It was a hot, humid day and they could not fill the balloon, much less launch it. The unsympathetic crowd, having already suffered numerous delays, forced Miollan and Janinet to flee and tore their balloon to shreds. The mob destroyed "tents, tables, and chairs" before the police restored order. Miollan and Janinet became the target of countless satirical verses and prints denouncing them as incompetent. One wit pointed out that the Abbé Miollan's name was an anagram for "balon abîmé" [sunken balloon].[57] His name became so well-known and so notorious that an opera patron used it to deal with a difficult situation. Apparently, a man attended the ballet and took his place on the parterre when a tall abbé stood in front of him, blocking his view. One of his companions announced to the other attendees that the offending abbé was Miollan. Soon, a chant of "There's Abbé Miollan" filled the room and the tall man had to leave the theater.[58]

Balloon launches were prone to disorder. Since the science of ballooning was still in its infancy, problems frequently arose that could postpone or even halt an ascent. Delays could last anywhere from a few hours to a few days. Meanwhile, huge crowds of people, many of whom took time off from work and traveled long distances, were gathered together with nothing to do but wait and, when the mood struck them, cause trouble. There were so many people who wanted to see the balloons that on at least one occasion people were asked to stay away so that actual scientific work could be performed. The Montgolfier brothers hired guards to protect their balloons.[59] The Parisian police lieutenant Lenoir ordered Miollan and Janinet to attempt their launch on a Sunday to prevent workers from taking time off and to keep employers happy. He made this decision after calculating that the sum of lost wages on a workday could surpass one hundred thousand *écus*.[60] Problems could also come from the balloon itself. Hot-air balloons stayed aloft because they were equipped with a small fire, typically burning straw or wool, to keep the air warm. Balloons, however, could not be steered and so tended to land on their own accord and without any regard to what happened to be on the ground in that location. Having a fire floating above cities, built largely of wood, was inherently problematic. For these reasons, state and local governments took an early interest in balloons.

A notice issued by the state in September 1783 announced the new invention in order to prevent the people from becoming alarmed by balloons in the sky. Such notices eventually led to attempts by the state to control balloon launches. In early 1784, the local government in Caen issued an ordinance demanding that all balloonists first obtain permission from the government. Lyon banned hot-air balloons and the *prévôt des marchands* in Lille requested a similar law the same day a balloon caught fire in that city. In Dijon the police forbade the making or launching of any balloons that included fireworks, flammable liquids, and other combustible items among the cargo. In Paris and Angers, failure to obtain prior permission for a balloon launch could result in a fine of five hundred livres.[61]

The question of utility

Balloons drew the attention of the general public, savants, the press, the court, and the police. It is possible, however, that the very ubiquity of balloons also served to bring their utility into question. While balloons were undoubtedly entertaining and marketable, many commentators doubted their usefulness to society. Critics faced an uphill battle due to the overwhelming popularity of balloons among savants and the general population alike. As one author noted, "the enthusiasm [for balloons] is so unlimited, that any men who dare to speak against the flying carriages are treated as ignorant, the enemies of useful science, of reason, and of good sense, etc."[62] The question of utility was an important one. But people asked it more quickly than they might have if only because it seemed that balloons were in the hands of popularizers and showmen rather exclusively controlled by savants. Almost immediately after the Montgolfier brothers launched their first balloon, people questioned its potential role in a variety of areas. As Benjamin Franklin quipped, when asked if he thought balloons were useful, "what use is a newborn baby?"[63] Some skepticism was well deserved. Balloons could not be controlled and simply sailed in whatever direction the wind might happen to be blowing. Nonetheless, the question of utility was, at least partly, a question of who should control this new invention.

One of the earliest descriptions of the Montgolfier brothers, that of Faujas de Saint-Fond describing their first launch in Annonay, began by claiming that Etienne and Joseph Montgolfier were "born with a taste for useful knowledge." One man sent an open letter to Faujas de Saint-Fond after a launch in which he claimed that balloons would provide for the French the "best education" thanks to their ability to "penetrate the atmosphere." In October 1783, a proponent of balloons wrote that they had "an incontestable utility."[64] While some people sought to provide specific examples of how balloons could be put to use, others claimed that they did not have to be defended at all. A lawyer from Guyenne wrote a letter to the editor of the *Affiches de province* saying that writers who questioned the public taste for balloons were really going about it the wrong way. What is the utility of fireworks, he asked, or of fountains in gardens or of the opera? He suggested that instead of discouraging scientific emulation, it would be more useful to allow people to experiment with balloons so that, maybe, some day specific applications might be found. For the time being, however, he argued that people should simply let the fashion for ballooning run its course.[65]

The ideas put forward for the possible uses of balloons ran the gamut from the ridiculous to the sublime, the practical to the impossible. Some contemporaries argued for the utility of balloons specifically within the sciences; it was difficult for them to conceive of anything so stunning as useless.[66] An anonymous letter published in the *Journal de Paris* less than three months after the first balloon launch in Paris criticized those individuals who had already written that balloons served no useful purpose. "It is proven," the author wrote, "that this discovery may have a lot of utility and is of no danger." This writer covers four areas in particular that may receive concrete benefits from the invention of balloons. In physics the author believed that analyses of gravity, the properties of the atmosphere, and electricity could all be performed in a balloon. In the area of astronomy, balloonists might study refraction and observe the horizon. Geographers could examine mountain ranges and measure the size of forests or the exact angles between specific locations more efficiently than before. Last, the author points to mechanics for even more practical utility. Balloons could be used, for example, to haul goods to a fort or city that was cut off by snow. Specifically, a fixed pulley could raise and lower objects from the balloon as needed. The author notes that this last use could come in handy in times of war. Furthermore, balloons might someday transport mail.[67]

Other commentators, writing in books and periodicals, provided additional examples of how science could take advantage of balloons. An amateur savant, the abbé des Granges Gagnière, noted that experiments with barometers at different heights could be made very easily with balloons. Lalande suggested that balloons could be used to measure longitude at sea. Another author, Arnaud de Saint-Maurice, called balloonists the "children of Newton" and provided a list of instruments and equipment that aerial navigators might take with them. Compasses and barometers topped the list but he also encouraged experiments with pendulums, magnets, and electricity.[68] Henri Decremps suggested that balloons could be useful for exploring and, more specifically, for demonstrating to native populations the superiority of the French. In this way, balloons became a potential tool for imperialism.[69]

While some people focused on the scientific utility of ballooning, others were quick to point out the possibilities for the military. The army could make great use of balloons for reconnaissance. Lookouts in balloons could tell the soldiers on the ground which way to go to either avoid or engage the enemy, depending on circumstances, as well as what was happening inside a walled city. Signals between balloons could also be used to synchronize attacks. In 1797 a physicist named Thilorier offered to construct a huge hot-air balloon capable of transporting an army into the heart of England. The French Revolutionary government even created the Compagnie aérostatique, a unit of which accompanied Napoleon to Egypt.[70] Another author also thought balloons could be used for moving troops. He imagined that if they could be enlarged, a mere twenty balloons might move as many as two thousand men and their munitions.[71]

Balloons also had great potential for commerce as a means of transporting goods. Etienne Montgolfier argued that ballooning must be retrieved from those individuals making it a spectacle and transformed into a commercially viable mode of transport. To this end he applied to the government for funds to perfect a method of steering balloons that would open up a trade route between Paris and the south of France. He believed that a balloon could be constructed capable of carrying a payload of forty thousand pounds and that sixteen people would be able to fly in it. After some negotiations, the state gave Etienne 40,000 livres to advance his project. More generally, balloons were seen as useful for reduced the weight of wagons and for transporting people. Simon-Nicolas-Henri Linguet suggested, somewhat tongue-in-cheek, that Parisian traffic jams might be eliminated thanks to balloons. On festival days, Linguet imagined, a thousand balloons of all different colors and shapes would fill the sky.[72] Not all commercial possibilities, it turns out, were of the positive kind. Balloons could also be used for illegal commerce, it was noted; thieving and kidnapping would be made considerably easier with the expediency of an escape route provided by a hot-air balloon. Thieves also took advantage of the crowds that developed around balloon launches to pick pockets; reportedly, Edmund Burke lost his purse and watch at such an event. The actions of lovers would also be made easier thanks to balloons; fathers would have a hard time preventing unwanted attentions to their daughters.[73]

The question of utility would have been easier to solve if someone had managed to invent a means by which balloons could be steered. Blanchard, as noted above, always claimed that he could navigate in spite of all evidence to the contrary. The Académie des Sciences, Belles-Lettres & Arts in Lyon established a prize of 1,200 livres to anyone who could indicate a sure method of steering balloons. Many people, savants and amateurs, put forward schemes to accomplish this. Some looked like oars for rowing in the ocean of air while others imitated the wings of birds; but none worked any better than the other. Despite multiple failures, savants continued to believe aerial navigation possible. Several savants penned pamphlets putting forward their ideas on the matter. No one, however, actually succeeded in developing a working method of steering a balloon.[74]

One of the problems inherent in the question of utility was the obvious fact that balloons, like popular science more generally, could easily be seen as both instructive and entertaining. The lines between entertainment and science crossed paths even for the balloonists themselves. While many claimed to perform balloon launches for the sake of science, and some actually did, it was clear that this was not always the case. Descriptions of the contents in balloons include, of course, a variety of scientific instruments but other things as well. Wine frequently made the trip in case the aeronauts needed some refreshment en route. A balloon launched from the Champs-Elysées in 1791 included an anchor, a compass, heaps of copies of the new constitution, a bit of bread, a bottle of wine, and two chicken legs. When Charles made his first flight, he left equipped with a barometer, thermometer, telescope, maps, and pencil and paper, as well as bottles of champagne, furs, and blankets. For Jean-Pierre Blanchard's forty-fifth flight, and the first in the United States of America, he "strengthened his stomach with a morsel of biscuit and a glass of wine." In addition, his friends surprised him by including a bottle of ether, a few drops of which "refreshed [him] very much" after he had achieved his optimal altitude.[75] Wine appears to have been fairly common in balloons. When the Robert brothers ascended in a balloon with the duc de Chartres and another, unnamed passenger, they brought with them six bottles, one of which broke due to an overly rapid descent. Not everybody was so lucky. When Richard

Crosbie tried to cross the Irish sea, from Ireland to England, in 1785, he crashed into the water. He had designed his gondola to float so he could stay above water until rescued, but he was much dismayed to learn that his bottle of cordial had broken in the "landing" thus depriving him of any refreshment while he waited for help.[76]

By 1787 the question of utility was still being raised but it was much harder to make a positive case. Piroux, the lieutenant of police in Nancy, wrote a letter to the *Journal de Nancy* in which he noted that it had been "four years and ten days since the balloon had been the object of pleasantries and of admiration in Europe." He noted that he was still waiting to hear an answer to the question "what use do balloons serve?" He admitted that the first balloons, those launched in Paris, had their usefulness. They were used to examine the stars more closely and to perform experiments on the density of the air and on how vision worked. Since that time, however, Piroux felt that balloons had become the purview of opportunists like Blanchard.[77] As another commentator put it, if one uses the same kind of reasoning as did Doctor Pangloss, then ballooning was "a good from which all the globists profit."[78] The benefits to the rest of the world, however, were less obvious.

Public interest and participation in this new invention rose quickly even as some savants decried ballooning as a useless activity, made all the more useless due to public involvement. Unfortunately, balloons, while regarded as emblematic of the power of the Enlightenment, seemed to serve no obvious function. Thus, although they were interesting and entertaining they were not put to much use. This is in contrast to opinion on divining rods that, despite their enormous utility in finding water and ore deposits, continued to have a disputed scientific legitimacy among savants even as the general public supported them. Nonetheless, the general public deemed balloon launches both scientifically instructive and entertaining.

Notes

1 For the "sublime invention" see Louis Sébastien Mercier and R. de la Brettone, *Paris le Jour, Paris la Nuit*, ed. Michel Delon and Daniel Baruch (Paris: Robert Laffont, 1990), 1051. On balloons as an emblem of the Enlightenment see Jean Sgard, "Les philosophes en montgolfière," *Studies on Voltaire and the Eighteenth Century* 303 (1992): 99–111. Francis Hopkinson to Thomas Jefferson, 31 March 1784, in Thomas Jefferson, *The Papers of Thomas Jefferson*, 28 vols. (Princeton: Princeton University Press, 1950–1995), VII:57.

2 For ballooning and inoculation see the *Discours sur les découvertes en général, et particulièrement sur deux des principales découvertes de ce siècle* (Paris: Philippe-Denys Pierres, 1784); on balloons and animal magnetism see Madame la Vicomtesse de Fars Fausse-Landry, *Mémoires de Madame la Vicomtesse de Fars Fausse-Landry, ou Souvenirs d'une octogénaire*, 2 vols. (Paris: Ledoyen, 1830), I:179–80. For the comparison between Louis XIV and Louis XVI and the list of immortals, see the anonymous *Lettre à M. de *** sur son projet de voyager avec la Sphère Aërostatique de M. de Montgolfier* (Paris: Marchands de Feuilles Volonts, n.d.), 3–4. For the "all important" discovery see *Courier de l'Europe*, 9 January 1784, 18.

3 The best book on ballooning in France is Charles C. Gillispie, *The Montgolfier Brothers and the*

Invention of Aviation, 1783–1784 (Princeton: Princeton University Press, 1983). More generally, see Marie Thébaud-Sorger, "'L'air du temps.' L'aérostation: savoirs et pratiques à la fin du XVIIIe siècle (1783–1785)," Thèse du doctorat (Ecole des hautes études en sciences socials, 2004); James Martin Hunn, "The Balloon Craze in France, 1783–1799: A Study in Popular Science," Ph.D. diss. (Vanderbilt, 1982). Also, see the comparative study by Richard Gillespie, "Ballooning in France and Britain, 1783–1786," *Isis* 75 (1984): 249–68.

4 For the fixed attention, *Procès-verbaux et détails des deux voyages aériens faits d'après la découverte de MM. Montgolfier* (Bruxelles: Bailly, 1783); for the city and court see the *Correspondance littéraire, philosophique et critique*, 16 vols., ed. Maurice Tourneux (Paris: Garnier, 1877–1882), XIII:344.

5 *Journal de Paris*, 8 November 1783, 1283–4.

6 Louis Sébastien Mercier, *Mon Bonnet de Nuit*, ed. Jean-Claude Bonnet (Paris: Mercure de France, 1999), 569.

7 Emmanuel, duc de Croy, *Journal inédit du duc de Croy*, 4 vols., eds. Vicomte de Grouchy and Paul Cottin (Paris: Flammarion, 1906), IV:307; and the anonymous, *Lettre à Mr. M. de Saint-Just, sur le globe aërostatique de MM. Montgolfier* (Paris: Mérigot, Royez, 1784), 3.

8 Cyrano de Bergerac, *Voyages to the Moon and the Sun*, trans. Richard Aldington (orig. edn 1657; reprint: New York: The Orion Press, 1962). Awen A.M. Coley, "Followers of Daedalus: Science and Other Influences in the Tales of Flight in Eighteenth-Century French Literature," *Studies on Voltaire and the Eighteenth Century* 371 (1999): 81–173. For Blanchard see for example *Affiches de Paris*, 1 October 1782, 2282; *Affiches de province*, 4 September 1782, 143; and *Journal de Nancy*, 1782, 239–40. On the literature of flight see Jules Duhem, *Histoire des idées aéronautiques avant Montgolfier* (Paris: Fernand Sorlot, 1943); and Marjorie Hope Nicolson, *Voyages to the Moon* (New York: Macmillan, 1948).

9 Charles Gillispie points out that up to this time, hydrogen had been measured in cubic inches rather than cubic feet and had only been manufactured for laboratory use. Gillispie, *The Montgolfier Brothers*, 29.

10 Jean-Baptiste-Joseph Fourier, "Eloge historique de M. Charles," *Mémoires de l'Académie Royale des Sciences de l'Institut de France* 8 (1829): lxxiii–lxxxviii.

11 On the sheep see *Courrier d'Avignon*, 10 October 1783. On the dangers of flying and the use of a worker see, *Courrier d'Avignon*, 12 September 1783, 292; and *Courrier d'Avignon*, 10 October 1783.

12 Pilâtre de Rozier actually went up in a tethered balloon as early as 15 October 1783, over a month before his free flight with d'Arlandes. Etienne Montgolfier, however, had probably gone up even a few days before this. Gillispie, *The Montgolfier Brothers*, 45.

13 On ballooning, and science more generally, in Saint Domingue, see James E. McClellan, *Colonialism and Science: Saint Domingue in the Old Regime* (Baltimore: Johns Hopkins University Press, 1992), 169–70. On ballooning in Oaxaca, see Virginia González Claverán, "Globos aerostáticos en la Oaxaca del siglo XVIII," *Quipu* 4 (1987): 387–400. On English ballooning see John Edward Hodgson, *The History of Aeronautics in Great Britain* (London: Oxford University Press, 1924); and Gillespie, "Ballooning in France and Britain, 1783–1786," 249–68. On Ireland see Barbara Traxler Brown, "French Scientific Innovation in Late-Eighteenth-Century Dublin: The Hydrogen Balloon Experiments of Richard Crosbie (1783–1785)," in *Ireland and the French Enlightenment, 1700–1800*, eds. Graham Gargett and Geraldine Sheridan (New York: St. Martin's Press, 1999): 107–26; and Linde Lunney, "The Celebrated Mr. Dinwiddie: An Eighteenth-Century Scientist in Ireland," *Eighteenth-Century Ireland* 3 (1988): 69–83.

14 On the efforts of the Montgolfier brothers to gain state patronage, see Gillispie, *The Montgolfier Brothers*, passim.

15 On the launch in Rodez, see *Affiches de Toulouse et du Haut-Languedoc*, 18 August 1784, 136.

16 For the subscription goals see Mathon de la Cour, "Lettre de Mathon de la Cour, Directeur de l'académie des sciences de Lyon, sur l'expérience de l'Aérostate que M. de Montgolfier a fait élever à Lyon," in Barthélemy Faujas de Saint-Fond, *Description des experiences*

aérostatiques de MM. De Montgolfier, 2 vols. (orig. edn, 1784; Osnabrück: Otto Zeller, 1968), II:84–5. For a list of the subscribers see the "Etat des souscriptions pour la grande experience du ballon aérostatique," in Raoul de Cazenove, *Premiers voyages aériens à Lyon en 1784* (Lyon: Pitrat, 1887), 60–63. On the social status of the subscribers and the satire of Mlle Salade see Hunn, "The Balloon Craze in France," 219–20.

17 A succinct account of this launch can be found in Gillispie, *The Montgolfier Brothers*, 72–79. On the lengthy list of amateurs who wanted to go up in the balloon, see *Journal de Paris*, 15 January 1784, 70–71.

18 *Affiches de province*, 3 November 1783, 1264–5.

19 *Affiches de province*, 20 March 1784, 171–2.

20 *Courrier d'Avignon*, 9 April 1784, 119–20. On Charles and the Robert brothers' tactics see Gillispie, *The Montgolfier Brothers*, 57. On the Robert brothers see Anne-Jean Robert and Nicolas-Louis Robert, *Mémoire sur les expériences aérostatiques* (Paris: L'Imprimerie de P.-D. Pierres, 1784). On the success of their subscription see Hunn, "The Balloon Craze in France," 136–7. Hunn cites the commentator Simeon-Prosper Hardy, *Mes loisirs*, 8 vols., Bibliothèque Nationale, Ancien fond français, 6680–7, citation in 6684, p. 390.

21 Gillispie, *The Montgolfier Brothers*, 31.

22 *Journal de Paris*, 19 September 1784, 1111 (Robert brothers); *Journal de Nancy*, 17 December 1783, 47–48 (Nicolas); and *Affiches des Trois-Evêches*, 25 January 1787, 26 (Brun); Alban and Vallet, *Précis des expériences faites par MM. ALBAN & VALLET; & Souscription proposée pour un Cours de Direction Aérostatique* (Paris: Velade, 1785).

23 On Tétu and his attempt to stay up in a balloon for twenty-four hours see for example *Affiches de province*, 24 June 1786, 298–9.

24 Gillispie, *The Montgolfier Brothers*, 27 (Charles); *Courrier d'Avignon*, 9 September 1783, 288 (on subscribers at the Café du Caveau); *Journal de Nancy*, 17 December 1783, 47–8 (Nicolas); and *Affiches des Trois-Evêches*, 25 January 1787, 26 (Brun).

25 *Réimpression de l'Ancien Moniteur* 5 (13 July 1790), 112.

26 The exact number of his ascents is debatable but was around sixty. On the number of people involved in balloon launches in Paris and the provinces see Hunn, "The Balloon Craze in France," 214–17. For his nickname see *Biographie nouvelle des contemporains*, 20 vols., eds. A.V. Arnault, et al (Paris: Librairie Historique, 1820–1825), III:52. On the accusation of greed see *Affiches de province*, 30 October 1784, 615.

27 Jean François Thomas Goulard, *Cassandre mécanicien, ou le bateau volant* (Paris: Brunet, 1783).

28 On the Académie Aérostatique see Léon Coutil, *Jean-Pierre Blanchard, physicien-aéronaute* (Evreux: Herissey, 1911), 13.

29 On his companions see Blanchard, *Liste chronologique des ascensions de l'Aéronaute Blanchard* (N.p., n.d.). This only goes up to 1803. For the parachuting dogs see Coutil, *Jean-Pierre Blanchard*, 13–16. On Garnerin, see Michel Poniatowski, *Garnerin: le premier parachutiste de l'histoire* (Paris: Albin Michel, 1983). On Garnerin and his dog see *Chronique de Paris*, 24 June 1790, 700.

30 On the incident with Dupont de Chambon see *Affiches de province*, 6 March 1784, 136; and Johann-Georg Wille, *Mémoires et journal de J.-G. Wille*, ed. Georges Duplessis, 2 vols. (Paris: Renouard, 1857), II:84–5.

31 On Marie-Madeleine-Sophie Blanchard see Poterlet, *Notice sur Madame Blanchard, aéronaute* (Paris: Imprimerie de Fain, 1819); Coutil, *Jean-Pierre Blanchard*, 20–24; and Rachel R. Schneider, "Star Balloonist of Europe: The Career of Marie-Madeleine Blanchard," *Consortium on Revolutionary Europe, 1750–1850: Proceedings* (1983): 697–711.

32 On Campmas see, e.g., *Affiches des Trois-Evêches*, 27 December 1787, 410; *Affiches des Paris*, 7 March 1789, 622; and *Journal de Paris*, 2 October 1784, 1167. On Enslen, see *Affiches des Trois-Evêches*, 25 January 1787, 30–1; *Affiches de Toulouse et du Haut-Languedoc*, 15 September 1784, 152; and *Affiches de Paris*, 8 September 1785, 2413–14.

33 For Enslen and Ruggieri, see *Affiches de Paris*, 18 October 1785, 2790–1; and *Journal de Paris*, 23 October 1785, 1219–20. For a somewhat lengthier description that mentions "all Paris"

had seen the balloons in advance, see *l'Année Littéraire*, VII (1785), 205–8. On the Jardin Ruggieri see Robert Isherwood, *Farce and Fantasy: Popular Entertainment in Eighteenth-Century Paris* (Oxford: Oxford University Press, 1986), 206–8.

34 *Journal de Paris*, 28 March 1789, 396 and also 11 April 1789, 458. See Hunn, "The Balloon Craze in France," 249, fn. 12.

35 *Courrier d'Avignon*, 23 September 1783, 303 (for the sale of toy balloons and the skirt); Jean-Claude Pingeron, *L'Art de faire soi-même les ballons aérostatiques, conformes à ceux de M. de Montgolfier* (Paris: Hardouin, 1783), 20–1 (on the prices of balloons for sale); *Courrier d'Avignon*, 21 October 1783 (for the hats and ribbons); *Affiches de Toulouse et du Haut-Languedoc*, 3 February 1784 (for hairstyles and buckles). On the hat worn by Madame Royale, see Monique de Huertas, *Madame Royale: L'énigmatique destinée de la fille de Louis XVI* (Paris: Pygmalion, 1999), 29. Hunn, "The Balloon Craze in France," 151–2 (for the household items, game, liquor, restaurant fare, and wax figure). For the presence of balloons on plates and furniture see Paul Guth, "Tous les arts se sont mis 'au ballon,'" in *Connaissance des Arts* 59 (January 1957): 14–19. As early as 1786 one commentator claimed that the fashion for balloons was already passing. See *Affiches de province*, 20 April 1786, 187–8. The Musée de l'air et de l'espace, at Le Bourget just outside of Paris, has a marvelous collection of balloon-related objects.

36 *Affiches de Paris*, 27 March 1784, 185 (for sale of balloons). For Bienvenu's course on gases see, for example, *Journal de Paris*, 1 May 1786, 491.

37 *Affiches de Paris*, 8 May 1791, 1728.

38 *Journal de Paris*, 27 November 1783, 1360; and *Affiches de Paris*, 27 November 1783, 2842.

39 *Journal de Paris*, 3 September 1783, 1014.

40 "Expériences faites à Paris," in Faujas de Saint-Fond, *Description des expériences*, I:272–3.

41 On the music, see *Affiches de la Province de Bretagne*, no. 25, no date, 227; for the cannon fire, see *Journal de Paris*, 28 August 1783, 991.

42 "Expérience faite à Versailles," in Faujas de Saint-Fond, *Description des expériences*, I:39–40; on the applause for Charles see *Journal de Paris*, 28 August 1783, 991.

43 On the emotion of the spectators, see the anonymous pamphlet, *Supplément à l'art de voyager dans les airs, contenant le Précis historique de la grande Expérience faite à Lyon le 19 Janvier 1784 et l'Exposé d'un moyen ingénieux pour diriger à volonté les Ballons aérostatiques* (N.p., n.d.), 17.

44 On the silence see "Expérience des Tuileries, second voyage aérien," in Faujas de Saint-Fond, *Description des expériences*, II:43. For the kneeling and fainting see "Lettre de M. Pilâtre de Rozier à M. Faujas de Saint-Fond," in ibid. II:78. On the stupefied crowd see Baronne d'Oberkirch, *Mémoires de la Baronne d'Oberkirch*, ed. Suzanne Burkard (Paris: Mercure de France, 1970), 356.

45 *Ségur,* Memoirs and Recollections of Count Louis Philippe de Ségur, (orig., edn., 1825; New York: Arno Press, 1970), 31.

46 *Journal de Paris*, 5 November 1783, 1272.

47 "Expérience de la Muette," in Faujas de Saint-Fond, *Description des expériences*, II:17 (on Notre Dame); Hunn, "The Balloon Craze in France," 250–2 (on the cemetery).

48 *Almanach des Ballons, ou Globes Aérostatiques; Etrennes du jour physico-historique et chantantes* (Paris: Langlois, 1784), 24–5 (for women in the rain), 69 (for mania). Anna Francesca Cradock, *Journal de Madame Cradock: Voyage en France (1783–1786)*, trans. and ed. O. Delphin Balleyguier (Paris: Perrin, 1896), 10.

49 *Courrier d'Avignon*, 12 December 1783, 395 (on Charles at the Tuileries); *L'Année littéraire*, VI (1783), 137 (on Charles at the Champs de Mars); *Annales politiques, civiles et littéraires du dix-huitième siècle*, ed. Simon-Nicolas-Henri Linguet (London: 1777–1792), XI:298 (on Versailles). On mass culture in France for the nineteenth century see Vanessa R. Schwartz, *Spectacular Realities: Early Mass Culture in Fin-de-Siècle Paris* (Berkeley: University of California Press, 1998); more generally, see James Naremore and Patrick Brantlinger, eds., *Modernity and Mass Culture* (Bloomington: Indiana University Press, 1991).

50 Maurice Quinlan, "Balloons and the Awareness of a New Age," *Studies in Burke and His Time*

14 (1972–1973), 225.

51 On Charles see *Courrier d'Avignon*, 30 December 1783, 416. *Journal encyclopédique*, October 1783, 126 (for stones).

52 *Journal de Nancy*, 12:18 (1783), 96.

53 See the image in Gillispie, *The Montgolfier Brothers*, plate IV (between pages 12 and 13).

54 Philalete, *Lettre de l'hermite de Nivolet sur l'expérience aérostatique faite à Chambéry, le 22 Avril 1784* (N.p., 1784), 2 (on the immense crowd). *Benjamin Franklin on Balloons*, ed. W.K. Bixby (Saint Louis: Bixby, 1922), n.p. also see *Courrier d'Avignon*, 1 June 1784, 179–80.

55 On the crowd in the Place des Victoires see "Expérience faite à Paris au Champ de Mars," in Faujas de Saint-Fond, *Description des expériences*, I:15–16. The story of Charles and the rioters is found in several places. See Lazare Carnot, *Mémoires sur Lazare Carnot, 1793–1823*, 2 vols. (Paris: Hachette, 1907), I:253.

56 On Miolan's courses see, for example, *Affiches de Paris*, 7 December 1783, 2921–2; and *Journal de Paris*, 5 March 1783, 269.

57 For a brief account of this episode see Gillispie, *The Montgolfier Brothers*, 97–8. On the subscription see *Journal de Paris*, 26 February 1784, 257–8. On the mob see the account in John Lough, ed., *France on the Eve of Revolution: British Travellers' Observations* (London: Croom Helm, 1987), 225. For some satirical verse see *Courrier d'Avignon*, 10 August 1784, 260; and *Correspondance littéraire, philosophique et critique* XIV:15–16. On the anagram see Pierre-Jean-Baptiste Nougaret, *Tableau mouvant de Paris, ou Variétés amusantes*, 3 vols. (Paris: Duchesne, 1787), I:128–9.

58 This story is recounted in Nikolai M. Karamzin, *Letters of a Russian Traveler, 1789–1790*, trans. Florence Jonas (New York: Columbia University Press, 1957), 213–14.

59 "Expériences faites à Paris, rue de Montreuil," in Faujas de Saint-Fond, *Description des expériences*, I:270. On the guards hired by the Montgolfier brothers see Hunn, "The Balloon Craze in France," 99. For a comparison between the attempts to maintain order in Paris and Bordeaux, see Hunn, "The Balloon Craze in France," chapter 5.

60 Hunn, "The Balloon Craze in France," 241.

61 For the notice to the people about balloons see *Avertissement au Peuple, sur l'Enlevement des Ballons ou Globes en l'air* (Paris: Herissant, 1783); Maurice Lantier, "Retombées aérostatiques sur la generalité de Caen (1783–1785)," *Annales de Normandie* 30 (1980), 272 (for Caen); Hunn, "The Balloon Craze in France," 236 (on Lyon); F. Isambert et al. (eds.), *Recueil général des anciennes lois françaises, depuis l'an 420 jusqu'à la revolution de 1789*, 29 vols. (Paris: Belin-Leprieur, 1822–1833), XXVII:403–4 (on flammable objects in balloons); *Ordonnance de Police, qui fait défenses de fabriquer & faire enlever des Ballons* (Paris: Pierres, 1784), 2; and *Ordonnance de Police, qui fait défenses à toutes personnes de quelque qualité & condition qu'elles soient, de fabriquer & faire enlever aucuns ballons* (Angers: Mame, 1784), 1 (on fines).

62 Fabry, *Réflexions sur la relation du voyage aérien de MM. Charles & Robert* (Paris: Chez les Libraires des Nouveautés, 1784), 4.

63 Gillispie, *The Montgolfier Brothers*, p. 52.

64 "Expérience faite à Annonay en Vivarais, le 5 Juin 1783," in Faujas de Saint-Fond, *Description des expériences*, I:1 (for the Montgolfier brothers' love of utility); "Lettre de M. Bourgeois, à M. Faujas de Saint-Fond," in Faujas de Saint-Fond, *Description des experiences*, I:263 (for the best education); "Lettre de M. Giroud de Villette," in Faujas de Saint-Fond, *Description des experiences*, I:280 (for the incontestable utility).

65 *Affiches de province*, 18 November 1784, 210–12.

66 Hunn, "The Balloon Craze in France," 172.

67 *Journal de Paris*, 17 November 1783, 1320–1, quote from page 1320.

68 For the abbé des Granges Gagnière see *Affiches de province*, 20 December 1783, 210–11; on Lalande see *Affiches de province*, 20 January 1784, 38–9. Arnaud de Saint-Maurice, *L'Observateur volant et le triomphe héroïque de la navigation aérienne, et des vésicatoires amusants et célestes, poëme en quatre chants* (Paris: Cussac, 1784), 63–4.

69 Henri Decremps, *La magie blanche dévoilée . . . [and] Supplément à la magie blanche dévoilée*, 2

parts in 1 vol. (Pairs: Langlois, 1784–1785), 119–32. Although bound together they are paginated separately; this comes from the supplement.

70 On balloons and the military see *Affiches de province*, 26 October 1783, 1231; and *Courrier d'Avignon*, 7 November 1783, 356. Patrice Bret, "Recherche scientifique, innovation technique et conception tactique d'une arme nouvelle: l'aérostation militaire (1793–1799)," in *Lazare Carnot, ou le savant-citoyen*, ed. Jean-Paul Charnay, et al. (Paris: Presses de l'Université de Paris, 1990): 429–51; Lucien Robineau, "Lazare Carnot et les compagnies d'aérostiers," *Revue historique des Armées* 2 (1989): 101–10; and Jacques Godechot, "Les premiers soldats du ciel: débuts de l'aérostation militaire," *Historiana* 11 (1985): 24–8. *Réimpression de l'Ancien Moniteur*, 29 (27 November 1797), 73 (on Thilorier).

71 On moving troops see, *Annales politiques, civiles et littéraires*, X:368.

72 *On Etienne Montgolfier's project see Gillispie,* The Montgolfier Brothers, 120–9; and Oscar Browning (ed), *Despatches from Paris, 1784–1790*, 2 vols. (London: Offices of the Society, 1909–1910), I:100–1. Also see "Lettre de M. de Saussure, à M. Faujas de Saint-Fond," in Faujas de Saint-Fond, *Description des expériences*, II:121. *Annales politiques, civiles et littéraires du dix-huitième siècle*, ed. Linguet (London: 1777–1792), X:367–8.

73 On thieves see *Annales politiques, civiles et littéraires*, X:365. On Edmund Burke, see Quinlan, "Balloons and the Awareness of a New Age," 227. On lovers see Hunn, "The Balloon Craze in France," 184–5.

74 On the prize, see *Journal de Nancy*, 12:18 (1783): 99–100. On the belief in the possibility of steering balloons see *Affiches de province*, 4 September 1783, 1022; and *Description de deux Machines propres à la Navigation aérienne* (N.p., n.d.), 1. Books on the topic include Jean-Louis Carra, *Essai sur la nautique aérienne* (Paris: Eugène Onfroy, 1784); and Felix Henin, *Mémoire sur la direction des aérostats* (Paris: Moreau, An X).

75 For the 1791 voyage see B. Lallemand de Sainte-Croix, *Procès-verbal très-intéressant du voyage aérien* (Paris: Imprimerie du Patriote François, 1791), 4. On Charles see Gillispie, *The Montgolfier Brothers*, 58. For Blanchard see Jean-Pierre Blanchard, *Journal of My Forty-Fifth Ascension and the First in America*, ed. Carroll Frey (Philadelphia: The Penn Mutual Life Insurance Company, 1943), 21–2.

76 *Gentleman's Magazine* 58 (1785): 652.

77 *Journal de Nancy*, 23:17 (1787): 4–47.

78 *Affiches de Toulouse et du Haut-Languedoc*, 28 January 1784, 16–17.

7
Conclusion

In 1790, J.-A.-C. Charles advertised several different lecture courses including two on electricity and one on optics. The ongoing political and cultural upheaval caused by the French Revolution did not slow down his business or hamper his ability to disseminate scientific ideas. In fact, Charles thrived during the 1790s. In addition to his usual courses he worked for the new government doing inventories of the *cabinets de physique* of émigrés who had fled the country after the onset of the revolution.[1] At the same time, when the new government created the Institut to replace the old royal academies, it rewarded Charles with a membership. In fact, from among the popularizers still working in France at the time of the Revolution, the Institut honored only Charles with inclusion in their new society. Charles's success demonstrates that popular science continued into the French Revolution; but science during that time period began to take on a new tone, with the professionalization of science education, a new focus on the utility of science for the state, and the emphasis on savants working directly for the nation. Popular science still existed, but a conscious divide was created between popular and elite science.

In many ways, popular science fit in quite well with the revolutionary spirit. The discourse of reason and utility that developed over the course of the eighteenth century, and that mid-level savants promoted in their courses, demonstrations, and clubs, created a sense of confidence in the abilities of the individual to conquer problems rationally and to solve seemingly impossible situations through an application of the same kind of reasoning taught at experimental physics courses. Just as members of the general public saw fit to offer their opinions of divining rods or support ballooning through subscriptions, so too did they feel they could tackle the job of shaping a new government. The appropriation of Enlightenment through public lecture courses of all kinds provided people with a sense of empowerment and entitlement. In any case, popularizers continued to disseminate science throughout the early years of the Revolution. The number of public lecture courses advertised in 1789 and 1790 in the *Affiches de Paris*, for example, maintained the same level as before the Revolution. Only the onset of the Terror brought an abatement of popular science advertisements. Arguably, however, this represented a rise in public interest for political news at the expense of scientific news, rather than a halt in the popularization of science.

Nonetheless, for reasons not necessarily connected to the Revolution, the position of popular science began to wane at the end of the eighteenth century. Starting in the mid-1780s, and influenced by the popularity of ballooning along with debates over issues like animal magnetism and divining rods, some savants began to attack science in the public sphere and demand its return to the halls of royal academies. These savants denigrated the role popular science had played during the Enlightenment. Innumerable courses taught during the eighteenth century had allowed men and women from all social classes and educational backgrounds to access the knowledge produced by savants throughout Europe. Public lectures enabled popularizers to create an oral *Encyclopédie* that reached those individuals who, either by inclination or circumstance, could not avail themselves of the volumes of Diderot and d'Alembert's project. Popular lectures addressed specific topics rather than discussing every aspect of human knowledge and so allowed individuals to pick and choose what they wanted to learn. At the same time, those lectures offered people disinclined or unable to read the opportunity to participate in the Enlightenment. Popular courses established themselves as an informal encyclopedia where interested people could go to learn, socialize, and earn cultural capital.

A significant difference existed, however, between the elite and the mid-level enterprise of scientific dissemination. Although at times elite science had entered the public sphere, for the most part the members of the Académie Royale des Sciences and other practitioners of elite science eschewed popular support and depended instead on their peers and the state for legitimacy. Mid-level savants, on the other hand, needed the approbation of the general public in order to survive economically. They required the approval of other savants and the state only if they desired to leave the realm of popular science and gain support through traditional old-regime methods. By taking science to the public, and by placing themselves in a position where their very livelihood depended on public support, popularizers became dependent on the audience that paid for their services. The members of the public sphere determined the nature of popularized science through its attendance at public lectures, or through public subscriptions in the case of balloons. At the same time, the audience members also influenced the location of popular science as well as its format and content.

Beginning in the 1730s and lasting well into the 1780s and 1790s, science formed a mainstay of French popular culture. The invention of ballooning in

1783 reinforced the belief that science could produce nearly miraculous discoveries. Indeed, faith in science went much further than many elite savants preferred. Once the public took over the sponsorship of balloons and began to speak with authority on divining rods, it seemed that anyone could claim to speak with scientific authority. Membership in a royal academy, or at least the approval of that academy, no longer seemed necessary; public approval on its own could legitimate the work of popularizers. Unfortunately, from the point of view of the elite, this left the door wide open for a broad range of rather dubious scientific efforts. The general public, however, had a firm belief in its ability to recognize and judge good science even as it took on entertaining and spectacular forms.

The greatest problems arose when the balance between usefulness and entertainment tipped too strongly in favor of the latter. Scientific showmen, emphasizing the theatrical side of science, mystified their work. This went in opposition to the idea that knowledge should help people avoid superstition and magic. Thus, just at the time that the popularizers had firmly established themselves in France, they became discredited in the eyes of their elite peers. At the same time, the avid participation of the public in supporting certain efforts, such as ballooning and divining rods, without distinguishing between them in terms of the quality of their theoretical underpinnings, further led savants to fret for the fate of science.

In addition to the controversy over divining rods, numerous other scientific debates filled the popular press and the cultural imagination of the French at this time. One charlatan raised over three thousand livres through a subscription for his "elastic shoes" with which he claimed he could walk on water. The use of a public subscription as well as the claim that he could conquer water, in a manner similar to that of the balloonists who conquered air, gave this experiment a solid, respectable status. Although exposed as a hoax, further claims to walking on water appeared over the next two years. [2] Another controversy developed over the efficacy of lightning rods. In 1780, Charles Dominique de Vissery de Bois-Valé, an amateur physicist residing in the town of St. Omer, attached a lightning rod to his house. His neighbors, however, expressed a deep concern over this experiment. They worried that a lightning rod might *attract* a lightning strike and cause a fire rather than draw lightning away from the buildings. The villagers in St. Omer took their case to court where, eventually, they faced Vissery's lawyer, a young man named Maximilien Robespierre. Vissery and Robespierre won their case in 1783.[3] The honor for the late eighteenth-century's most infamous scientific controversy fell on Franz-Anton Mesmer and his claim to cure people using animal magnetism. This topic provoked a heated debate among savants and amateurs alike.[4] All these debates drew on public as well as scientific support and relied heavily on both the credulity of the French and an unwavering belief in the power of science to overcome all obstacles.

Thanks to this ongoing mixture of science with mystery and spectacle, individuals wrote a number of books in the 1780s with the chief goal of exposing what lay behind the efforts of scientific popularizers, especially those with a particularly theatrical style of presentation. Henri Decremps, a writer and amateur savant, made it his mission in life to "unmask" many of the mid-level savants who practiced their trade in and around Paris. As we saw in the discussion of divining rods, he attacked dowsing with great abandon. However, he gave considerable attention to other popularizers, like Giuseppe Pinetti, who in his opinion tricked the audience into believing that something scientifically miraculous had occurred during his courses. Decremps wanted to "satisfy the curiosity of the intelligent reader" and help audiences "glimpse all of the internal workings put into play in order to amuse and seduce them" at popular science lectures.[5] Dedicating his work to "reason," Decremps tried to expose the methods used by those men who "pretended to do incomprehensible things" and who tried "to pass themselves off as magicians."[6] "Mankind," Decremps noted, was more easily "deceived than undeceived."[7] He was reacting against the use of spectacle and what he perceived to be a lack of utility in popular science, especially in courses found on the Boulevard du Temple. Instead of the way in which popular science had been practiced, Decremps advocated, during the French Revolution, the teaching of science to the sans-culottes, a proposal that never really got off the ground.[8]

Pinetti did not allow Decremps's challenge to go unanswered. In 1784 he set up a benefit performance of his physics lecture show, with the profits going to the poor, in which he clearly outlined the scientific nature of his experiments and demonstrated explicitly how he executed them. He satisfied the questions of his audience so thoroughly, claimed one contemporary, that even "one of the friends of [Decremps], present at the meeting, could not avoid sharing" Pinetti's triumph. For the most part, however, Decremps's efforts to debunk popular science met with praise.[9]

The combination of utility and spectacle that had worked so well in the eighteenth century led to a backlash in other quarters as well. Elite savants developed a distinction between their work and that of the popularizers. This division appears in the *Encyclopédie méthodique*, a successor to the *Encyclopédie* of Diderot and d'Alembert. It included both a multi-volume *Dictionnaire de physique* as well as volumes titled *Dictionnaire encyclopédique des amusemens des sciences mathématiques et physiques* and *Dictionnaire des jeux mathématiques*. Ostensibly, Jacques Lacombe wrote the latter two works, although for the most part he compiled them by borrowing liberally (usually acknowledging the source) from other works disseminating natural philosophy. Nollet's work received due notice alongside the contributions gleaned from Pinetti and Guyot. By drawing on these and other esteemed authors, Lacombe claimed "the useful [was] nearly always united with the agreeable." "There are, it is true, some games," he added, "but these games evolved chiefly from the results or the solutions drawn from the most abstract and subtle of the sciences and the arts."[10]

Lacombe's dictionaries mark a significant turning point regarding the dissemination of natural philosophy and should be viewed in sharp contrast to the more academic *Dictionnaire de physique*. While the latter work contains a wealth of information concerning the nature of mathematical physics, it also required a considerable level of learning to wade through some of the explanations. Lacombe's compilations, on the other hand, borrowed directly from the experimental natural philosophy that popularizers had addressed to a popular audience over the preceding century. Covering topics such as electricity, animal magnetism, automata, and talking statues, Lacombe tapped directly into the realm of popularized experimental physics that had grown so significantly over the eighteenth century. The different volumes of the *Encyclopédie méthodique* marked the divide between two realms of natural philosophy at the end of the Enlightenment; elite savants separated "real" science from what was, from their point of view, game playing as represented by the work of popularizers. Those individuals associated with the seemingly more frivolous style of physics found themselves relegated to the charlatanesque side of the Enlightenment.[11] The distance between state-sponsored, official science and the science recognized by the majority of the people grew wider.

In the end, the period of the French Revolution had something of an adverse effect on popular science. The revolutionary period saw an emphasis on the potential of science to serve the new, rational state; science education became a significant focus of political attention, as did the ability of savants to provide useful services to the nation.[12] Most popularizers could not meet these expectations. Charles, on the other hand, performed admirably due to his flexibility and willingness to work for the new government. As a result, he became a member of the Institut de France. It helped, perhaps, that the most popular of all sciences, ballooning, did find a place during the French Revolution in the form of the Compagnie aérostatique, a ballooning branch of the military. In any case, the expansion and professionalization of education during the Revolution initiated a downward spiral for popular science practitioners whose efforts were seen as trivial. As a result, the status of popularizers declined precipitously in the period of the French Revolution and after. By the nineteenth century, popularizers of science had lost much of their former status within the scientific community.

Popular science formed a major part of the public sphere during the eighteenth century. Many people appropriated science through public lecture courses and expressed considerable interest in scientific work through their membership in clubs and their participation in the funding of balloons and the debate over divining rods. Thus, the audience for popular science contributed their own ideas about natural philosophy back into the cultural pool. The consumption of ideas from the field of experimental physics enabled a significant subset of the French population to gain access to, and participate in, the cultural Enlightenment. The connections between mid-level and elite savants as well as the links between the public sphere and the Enlightenment project provide us with a very dynamic, social and economically more complete picture of the Enlightenment. Popularizers played an active social, economic, and intellectual role between the Enlightenment and the people. In spite of their ill treatment at the hands of the new group of professional scientists who emerged during the French Revolution, and their subsequent marginalization by historians, popularizers of science represented an important aspect of the dissemination of the Enlightenment and the growth of scientific culture.

Notes

1 For the courses Charles offered see *Affiches de Paris*, 24 April 1790, 1059; ibid., 5 October 1790, 3085; and ibid., 23 June 1790, 1812. On his work for the revolutionary government see AN, F17–1270.

2 On the "elastic shoes" see Robert Darnton, *Mesmerism and the End of the Enlightenment in France* (Cambridge, MA: Harvard University Press, 1968), 23–4; and *Affiches de province*, 18 December 1783, 1448; 27 December 1783, 223. For additional claims see *Affiches de Province*, 10 August 1784, 454; and ibid., 10 December 1785, 595.

3 On the trial over lightning rods see Jessica Riskin, *Science in the Age of Sensibility: The Sentimental Empiricists of the French Enlightenment* (Chicago: University of Chicago Press, 2002), 139–87.

4 On Mesmer see Darnton, *Mesmerism*; Riskin, *Science in the Age of Sensibility*, 189–225; Charles Gillispie, *Science and Polity in France at the End of the Old Regime* (Princeton: Princeton University Press, 1980), 261–89; Laurence Brockliss and Colin Jones, *The Medical World of Early Modern France* (Oxford: Clarendon, 1997), 794–802.

5 Henri Decremps, *La Magie blanche dévoilée*, [and] *Supplément à la magie blanche dévoilée*, 2 parts in 1 vol. (Paris: Langlois, 1784–1785), viii (quote from part one). Also see Decremps, *Testament de Jérôme Sharp, Professeur de Physique amusant* (Paris: Grander, 1786); Decremps, *Les Petites Aventures de Jérôme Sharp*, ed. Jean-Marc Drouin (Paris, 1789; Paris: Editions Champion Slatkine, 1989); and Descremps, *Codicile de Jérôme Sharp* (Paris: F.J. Desoer, 1791).

6 Decremps, *Supplément*, 13–14.

7 Decremps, *Supplément*, title page.

8 Riskin, *Science in the Age of Sensibility*, 254–68.

9 For Pinetti's benefit lecture see *Journal de Paris*, 13 March 1784, 325. For the positive views of Decremps see *Affiches de province*, 30 April 1785, 206; *Journal encyclopédique*, June 1785, 280–92; and Pierre-Jean-Baptiste Nougaret, *Tableau mouvant de Paris, ou, Variétés amusantes*, 3 vols. (Paris: Duchesne, 1787), II: 316–18, III: 355–8. The exception to this praise for Decremps can be found in the very negative entry, written by E. Jouy, found in the *Biographie nouvelle des contemporains*, ed. A.V. Arnault et al., 20 vols. (Paris: Librarie Historique, 1820–1825), V:262–3. For his response to this criticism see Decremps, *Lettre à M. Jouy* (Paris: Carilian-Goeury, 1825).

10 Lacombe, *Dictionnaire encyclopédique des amusemens des sciences mathématiques et physiques* (Paris: Panckoucke, 1792), vii–viii. See also Lacombe, *Dictionnaire des jeux mathématiques* (Paris: H. Agnasse, An VII); and Gaspard Monge et al., *Dictionnaire de physique*, 4 vols. (Paris: Hôtel de Thou, 1793–1822). On the *Encyclopédie méthodique* see Christabel P. Braunrot and Kathleen

Hardesty Doig, "The *Encyclopédie méthodique*: An Introduction," *Studies on Voltaire and the Eighteenth Century* 327 (1995): 1–152; George B. Watts, "The *Encyclopédie méthodique*," *Publications of the Modern Language Association* 73 (1958): 348–66; Suzanne Delorme and René Taton (eds.), *"L'Encyclopédie" et le progrès des sciences et des techniques* (Paris: Presses Universitaires de France, 1952); and Robert Darnton, *The Business of Enlightenment: A Publishing History of the Encyclopédie, 1775–1800* (Cambridge, MA: Belknap, 1979), 395–459.

11 Darnton, for example, calls Lacombe's dictionaries "frivolous" and places them alongside "light literature" rather than science. Darnton, *The Business of Enlightenment*, 435. On the professionalization of science during the French Revolution see Maurice Crosland, "The Development of a Professional Career in Science in France," *Minerva* 13 (1975): 38–57.

12 On science during the period of the Revolution see for example Charles C. Gillispie, *Science and Polity in France: The Revolutionary and Napoleonic Years* (Princeton: Princeton University Press, 2004); Nicole Dhombres and Jean Dhombres, *Naissance d'un pouvoir: Sciences et savants en France (1793–1824)* (Paris: Editions Payot, 1989); and Dorinda Outram, "The Ordeal of Vocation: The Paris Academy of Sciences and the Terror, 1793–95," *History of Science* 21 (1983): 251–73.

Select biblography

Archival material

Archives Nationales (AN)
Fonds Publics de l'Ancien Régime
Series O1: Maison du Roi
Series T: Papiers privés tombés dans le domaine public
Series Y: Châtelet de Paris et prévôté d'Ile-de-France
Fonds Publics Postérieurs à 1789
Series C: Assemblées nationales
Series F12: Versement des ministères et des administrations qui en dependent: Commerce et industrie
Series F17: Versement des ministères et des administrations qui en dependent: Instruction publique
Minutier central des notaires de Paris (MC)
Archives de l'Académie des Sciences (Institut de France): Dossiers on members
Archives de Paris: Administration des Domains, Ville de Paris, Lettres de Chancellerie
Bibliothèque de l'Institut (BI): Fonds Joseph Bertrand
Bibliothèque historique de la ville de Paris (BHVP)
Bibliothèque municipale d'Orléans: Mémoires de Lenoir, Ms. 1423.
Conservatoire nationale des arts et métiers
Musée de l'air et de l'espace, Le Bourget: Fonds Montgolfier

Primary printed sources (all titles left in original spelling)

Affiches d'Angers. 1773–1811.
Affiches d'Artois, le Boulonnois et le Calaisis. 1788–1789.
Affiches de Bordeaux. 1758–1784.
Affiches de Dauphiné. 1774–1792.
Affiches de la Basse-Normandie. 1786–1788.

Affiches de la Haute et Basse Normandie. 1762–1773.
Affiches de l'Orléannais. 1764–1790.
Affiches de Normandie. 1762–1791.
Affiches de Paris. Paris, 1745–1751. Becomes *Annonces, affiches et avis divers*. Paris, 1751–1782. Becomes *Affiches, annonces et avis divers*. Paris, 1783–1811.
Affiches de Picardie, Artois, Soissonnois et Pays-Bas François. 1773–1775, 1779–1780.
Affiches de Poitou. 1773–1789.
Affiches de province, ou Annonces, affiches et avis divers. Paris, 1752–1761. Becomes *Affiches, annonces et avis divers*. Paris, 1761–1784. Becomes *Journal Général de France*. Paris, 1785–1792.
Affiches de Toulouse et du Haut-Languedoc. 1781–1789.
Affiches des Trois-Evêches. 1771–1791.
Affiches du Beauvaisis. 1786–1788.
Affiches du Hainaut & du Cambresis. Also *Journal du Hainaut & du Cambresis*. 1788–1789.
Alban and Vallet. *Précis des expériences faites par MM. ALBAN & VALLET; & Souscription proposée pour un Cours de Direction Aérostatique*. Paris: Velade, 1785.
Almanach Dauphin, ou Tablettes royales de correspondance et d'indication générale. Paris, 1789.
Almanach des Ballons, ou Globes Aérostatiques; Etrennes du jour physico-historique et chantantes. Paris: Langlois, 1784.
Almanach forain. 7 vols. Paris, 1773–1787. Also known as *Les Spectacles des foires et des boulevards de Paris* and *Les Petits Spectacles de Paris*.
Almanach Parisien, en faveur des étrangers et des personnes curieuses . . . Paris: Duchesne, 1765–1793.
L'Amour physicien, ou l'origine des ballons. Paris: Cailleau, 1784.
Annales politiques, civiles et littéraires du dix-huitième siècle. Ed. Simon-Nicolas-Henri Linguet. London: 1777–1792.
L'Année littéraire, ou suite des lettres sur quelques écrits de ce temps. Ed. Elie Fréron. 56 vols. Amsterdam: C.J. Panckoucke, 1754–1790.
Annuaire du Lycée des Arts pour l'an III. Paris: Chez la Concierge du Lycée des Arts, An III.
Arbeltier, M. *Lycée des femmes. Plan de cet établissement*. Paris: L'Imprimerie de Roland, n.d.
Archives parlementaires de 1787 à 1860. Paris: Librairie administrative de Paul Dupont, 1787–.
d'Argenson, René de Voyer. *Journal et mémoires du marquis d'Argenson*. 9 vols. Ed. E.J.B. Rathery. Paris: Jules Renouard, 1859–1867.
d'Arlandes, François Laurent and Jacques-Alexandre-César Charles. *Les Voyageurs aëriens*. N.p., 1784.
Athénée de Paris. *Fondateurs*. Paris, 1808.
Avant-coureur, Feuille hebdomadaire. 14 vols. Paris, 1760–1773. Was *La Feuille nécessaire, contenant divers détails sur les sciences, les lettres et les arts*. Paris, 1759. Becomes *Gazette de Avant-coureur*. Paris, 1774.
Avertissement au Peuple, sur l'Enlevement des Ballons ou Globes en l'air. Paris: Herissant, 1783.

Avis divers: Supplement de Annonces, affiches et avis divers (Paris, 1751–1782). 5 vols. Paris, 1777–1778.

Bachaumont, Louis Petit de. *Mémoires secrets pour server à l'histoire de la République des Lettres en France depuis 1762 jusqu'à nos jours*. 36 vols. London: John Adamson, 1784–1789.

[Bassi?] *Lycée de Paris*. Paris, 1784.

Bayle, Pierre. *Dictionnaire historique et critique*. 16 vols. Geneva: Slatkine, 1969.

Bibliothèque physico-économique, instructive et amusant. Paris, 1782–1798.

Blanchard, Jean-Pierre. *Journal of My Forty-Fifth Ascension and the First in America*. Ed. Carroll Frey. Philadelphia: The Penn Mutual Life Insurance Company, 1943.

———. *Liste chronologique des ascensions de l'Aéronaute Blanchard*. N.p., n.d.

Brisson, Mathurin-Jacques. *Dictionnaire raisonné de physique*. 3 vols. Paris: Leboucher & Lamy, 1781.

———. *Observations sur les nouvelles découvertes aërostatiques*. Paris: Boucher & Lamy, 1784.

Brissot de Warville, Jacques-Pierre. *Mémoires de Brissot*. 4 vols. Ed. M.F. de Montrol. Paris: Ladvocat, 1830.

Browning, Oscar, (ed.). *Despatches from Paris, 1784–1790*. 2 vols. London: Offices of the Society, 1909–1910.

Buchoz, Pierre-Joseph. *La Nature considerée sous des différens aspects, ou Journal des Trois regnes de la nature*, 5 vols. Paris, 1780–1783.

Bugge, Thomas. *Science in France in the Revolutionary Era*. Ed. Maurice Crosland. Cambridge, MA: The M.I.T. Press, 1969.

Bussière, Paul. *Lettre à M. l'abbé D.L.****. Paris: Louis Lucas, 1694.

Carra, Jean-Louis. *Essai sur la nautique aérienne*. Paris: Eugène Onfroy, 1784.

Chauvin, Pierre. *Lettre à Madame la Marquise de Senozan*. In Pierre Lebrun, *Histoire critique des pratiques superstitieuses*. 4 vols. 2nd edn, augmenté. Amsterdam: Jean-Frederic Bernard, 1733–1736: III:1–28.

Chronique de Paris. Paris, 1789–1793.

Comiers, Claude. *La Baguette justifée, et ses effets démontrez* naturels. N.p., 1693.

———. *Factum pour la baguette divinatoire*, N.p., 1693.

Conseil du premier musée. (N.p., n.d.).

Correspondance littéraire, philosophique et critique. 16 vols. Ed. Maurice Tourneux. Paris: Garnier frères, 1877–1882.

Courier de l'Europe. London: Cox, 1773–1793.

Courrier d'Avignon. 1733–1792.

Cradock, Anna Francesca. *Journal de Madame Cradock: Voyage en France (1783–1786)*. Trans. and ed. O. Delphin Balleyguier. Paris: Perrin, 1896.

Croy, Emmanuel duc de. *Journal inédit du duc de Croy, 1718–1784*. 4 vols. Eds. Vicomte de Grouchy and Paul Cottin. Paris: Flammarion, 1906.

Decremps, Henri. *Codicile de Jérôme Sharp, professeur de physique amusante*. Paris: F.J. Desoer, 1791.

———. *Lettre à M. Jouy*. Paris: Carilian-Goeury, 1825.

———. *La Magie blanche dévoilée*. [and] *Supplément à la magie blanche dévoilée*. 2 parts in 1 vol. Paris: Langlois, 1784–1785.

———. *Les Petites Aventures de Jérôme Sharp, professeur de physique amusante*. Ed. Jean-Marc Drouin. 1789. Reprint. Paris: Editions Champion-Slatkine, 1989.

———. *La Science sanculottisée*. Paris: Chez l'Auteur, 1793–1794.

———. *Testament de Jérôme Sharp*. Paris: Grander, 1786.

Deparcieux, Antoine. *Dissertation sur les globes aérostatiques*. Paris: Chez l'Auteur, 1783.

Description de deux Machines propres à la Navigation aérienne. N.p., n.d.

[Devilliers, Ainé.] *Les Numéros parisiens, ouvrage utile et nécessaire aux voyageurs à Paris*. Paris: L'Imprimerie de la vérité, 1788.

Diderot, Denis. "The Indiscreet Jewels." Trans. Sophie Hawkes. In *The Libertine Reader: Eroticism and Enlightenment in Eighteenth-Century France*. Ed. Michel Feher. New York: Zone, 1997: 333–541.

———. *Lettres à Sophie Volland*. 3 vols. Ed. André Babelon. Paris: Gallimard, 1930.

Discours sur les découvertes en général, et particulièrement sur deux des principales découvertes de ce siècle. Paris: Philippe-Denys Pierres, 1784.

Dufieu, Jean Ferapie. *Manuel physique, ou manière courte et facile d'expliquer les phénomènes de la nature*. Paris: J.-T. Herissant, 1758.

Dulaure, Jacques-Antoine. *Nouvelles descriptions des curiosités de Paris*. 2 vols. Paris: Lejay, 1785.

Duplessy, Mme la Baronne. *Répertoire des lectures faites au Musée des Dames*. Paris: Cailleau, 1788.

Fabry, *Réflexions sur la relation du voyage aérien de MM. Charles & Robert et la brochure intitulée: Méthode aisée de faire la Machine aérostatique*. Paris: Chez les Libraires des Nouveautés, 1784.

Famin, Pierre-Noel. *Cours abrégé de physique expérimentale, à la portée de tout le monde*. Paris: Briand, 1791.

Faujas de Saint-Fond, Barthélemy, ed. *Description des expériences aérostatiques de MM. De Montgolfier*. 2 vols. 1784. Reprint. Osnabrück: Otto Zeller, 1968.

Fontenelle, Bernard le Bovier de. *Conversations on the Plurality of Worlds*. Trans. H.A. Hargreaves. Berkeley: University of California Press, 1990.

Fourier, Jean-Baptiste-Joseph, le Baron. "Eloge historique de M. Charles." *Mémoires de l'Académie Royale des Sciences de l'Institut de France* 8 (1829): lxxiii–lxxxviii.

Franklin, Benjamin. *Benjamin Franklin on Balloons: A Letter Written from Passy, France, January Sixteenth MDCCLXXXIV*. Ed. W.K. Bixby. Saint Louis: W.K. Bixby, 1922.

Garnerin, André-Jacques. *Parc de Mousseaux. Voyage aérien du Citoyen Garnerin, avec une jeune personne*. Paris: Imprimerie du Cercle-Social, an VI.

Garnier, Pierre. *Dissertation physique en forme de lettre de Seve, Seigneur de Flecheres*. In Pierre Lebrun, *Histoire critique des pratiques superstitieuses*. 4 vols. 2nd edn, augmentée. Amsterdam: Jean-Frederic Bernard, 1733–1736: III:29–63.

———. *Histoire de la baguette de Jacques Aymar*. Paris: Jean-Baptiste Langlois, 1693.

Genlis, Stephanie Félicité Ducrest de Saint-Aubin, comtesse de. *Mémoires inédits de Madame la Comtesse de Genlis*. 10 vols. Paris: Ladvocat, 1825.

Goulard, Jean François Thomas. *Cassandre mécanicien, ou le bateau volant*. Paris: Brunet, 1783.

Grandjean de Fouchy, Jean-Paul. "Eloge de M. l'Abbé Nollet." *Histoire de l'Académie Royale des Sciences* (1770): 121–36.

Guyot, M. *Essai sur la construction des ballons aérostatiques et sur la manière de les diriger*. Paris: Gueffier, 1784.

——. *Nouvelles récréations physiques et mathématiques*. 4 vols. Paris: Gueffier, 1769–1770.

Hardy, Siméon-Prosper. *Mes loisirs*. Eds. Maurice Tourneux and Maurice Vitrac. Paris: Picard & fils, 1912.

Henin, Felix. *Mémoire sur la direction des aérostats*. Paris: Moreau, An X.

Histoire de l'Académie Royale des Sciences, avec les Mémoires de Mathématiques et de Physique. Paris, 1698–1790.

Histoire véritable et merveilleuse d'une jeune angloise, Précédée de quelques circonstances concernant l'Enfant Hydroscope, & de beaucoup d'autres traits & Phénomènes les plus singuliers dans ce genre. Paris: Lottin, 1772.

Isambert, F., et al., eds. *Recueil général des anciennes lois françaises, depuis l'an 420 jusqu'à la révolution de 1789*. 29 vols. Paris: Belin-Leprieur, 1822–1833.

Jèze, M. *Etat ou Tableau de la ville de Paris*. Paris: Prault, 1763.

Journal de Monsieur, frère du roi. Paris, 1778–1783.

Journal de Nancy. 1778–1787.

Journal de Paris. Paris, 1777–1800.

Journal de physique. Eds. François Rozier, Jean-André Mongez, and Jean-Claude Delametherie. Paris, 1773–1793.

Journal de politique et de littérature. Ed. Simon-Nicolas-Henri Linguet. Bruxelles, 1774–1778.

Journal de Trévoux. Paris, 1701–1767.

Journal des beaux-arts et des sciences. Paris, 1768–1774, 1776–1778.

Journal des sçavans. Paris, 1665–1797.

Journal des sciences utiles. Ed. Pierre Bertholon. Paris: Périsse, 1791.

Journal du Lycée des Arts. Paris, 1793.

Journal encyclopédique. Paris, 1756–1793.

Karamzin, Nikolai M. *Letters of a Russian Traveler, 1789–1790*. Trans. Florence Jonas. New York: Columbia University Press, 1957.

Lacombe, Jacques. *Dictionnaire des jeux mathématiques*. Paris: H. Agnasse, An VII.

——. *Dictionnaire encyclopédique des amusemens des sciences mathématiques et physiques*. Paris: Panckoucke, 1792.

Lallemand de Sainte-Croix, B. *Procès-verbal très-intéressant du voyage aérien*. Paris: Imprimerie du Patriote François, 1791.

Launoy, and François Bienvenu. *Instruction sur la nouvelle Machine inventée par MM. Launoy, Naturaliste, & Bienvenu, Machiniste-Physicien*. N.p., n.d.

Lavoisier, Antoine-Laurent. *Oeuvres de Lavoisier*. 6 vols. Paris: Imprimerie Impériale, 1862–1893.

Lebrun, Pierre. *Histoire critique des pratiques superstitieuses*. Paris: Jean de Nully, 1702.

——. *Histoire critique des pratiques superstitieuses*, 4 vols. 2nd edn. Paris: Delaulne, 1732–1737.

——. *Histoire critique des pratiques superstitieuses*. 4 vols. 2nd edn, augmentée. Amsterdam: Jean-Frederic Bernard, 1733–1736.

——. *Lettres qui découvrent l'illusion des philosophes sur la baguette, et qui détruisent leurs systèmes*. Paris, 1693. Bound with Pierre le Lorraine de Vallemont, *La Physique Occulte, Ou Traité de la Baguette Divinatoire . . . Augumentée de Plusieurs Pieces*. Paris: Jean Coudot, 1696.

Ledru, Nicolas-Philippe. *Rapport de MM. Cosnier, Maloet, Darcet, Philip, le Preux, Desessartz, & Paulet, Docteurs-Régens de la Faculté de Médecine de Paris*. Paris: Philippe-Denys Pierres, 1783.

Lemaistre, John G. *A Rough Sketch of Modern Paris, or, Letters on Society, Manners, Public Curiosities, and Amusements in that Capital*. 2nd edn. London: Printed for J. Johnson, in St.-Paul's Churchyard, 1803.

*Lettre à M. de *** sur son projet de voyager avec la Sphere Aërostatique de M. de Montgolfier*. Paris: Marchands de Feuilles Volonts, n.d.

Lettre à Mr. M. de Saint-Just, sur le globe aërostatique de MM. Montgolfier. Paris: Mérigot, Royez, 1784.

Liste de toutes les personnes qui composent le premier musée autorisé par le Gouvernement, sous la protection de Monsieur et de Madame. Pour l'Année 1785. Paris: Imprimé par Ordre de Conseil de Musée, 1785.

Lycée des Arts: Prospectus. Paris: Au local du Lycée des Arts, Jardin Egalité, An III.

Lycée républicain. Souscription pour l'an IX. Seizième année lycéenne. N.p., n.d.

Marat, Jean Paul. "Les Charlatans modernes." In *Les Pamphlets de Marat*. Ed. Charles Vellay. Paris: Charpentier & Fasquelle, 1911: 255–96.

——. *La Correspondance de Marat*. Ed. Charles Vellay. Paris: Charpentier & Fasquelle, 1908.

Mayeur de Saint-Paul, François-Marie. *Le chroniqueur désoeuvré, ou l'Espion du boulevard du Temple*. London, 1782.

——. *Tableau du nouveau Palais Royal*. 2 vols. London, 1788.

Mémoires de l'Académie Royale des Sciences. Paris, 1698–1790.

Mémoires de mathématiques et de physique. 9 vols. Paris: L'Imprimerie Royale, 1750–1786.

Mémoires du Musée de Paris. 2 vols. Paris: Moutard, 1784–1785.

Ménestrier, Claude-François. "Des indications de la baguette pour découvrir les sources d'Eau, les Métaux cachez, les Vols, les Bornes déplacées, les Assassinats, &c." In *La Philosophie des images énigmatiques*. Lyon: Hilaire Baritel, 1694.

Mercier, Louis Sébastien. *Les Entretiens du Palais Royal*. Utrecht: Buisson, 1786.

——. *Mon Bonnet de Nuit*. Ed. Jean-Claude Bonnet. Paris: Mercure de France, 1999.

——. *Panorama of Paris*. Ed. Jeremy D. Popkin. University Park: Penn State University Press, 1999.

—— and Restif de la Bretonne. *Paris le jour, Paris la nuit*. Eds. Michel Delon and Daniel Baruch. Paris: R. Laffont, 1990.

Mercure de France. Paris, 1724–1791.

Mercure galant. Paris, 1672–1710.

Metra, François, Guillaume Imbert de Boudeaux, and A.-B.-L. Grimod de La

Reynière. *Correspondance secret, politique et littéraire*. 16 vols. London: John Adamson, 1787–1789.

Miremont, Anne d'Aubourg. *Traité de l'éducation des femmes, et Cours complet d'instruction*. 7 vols. Paris: Ph.-D. Pierres, 1779–1789.

Monge, Gaspard, et al. *Dictionnaire de physique*. 4 vols. Paris: Hôtel de Thou, 1793–1822.

Moreau de Saint-Méry, Mederic-Louis-Elie. *Discours sur l'utilité des assemblées publiques littéraires*. Parma: Bodoni, 1805.

——. *Discours sur l'utilité du musée établi à Paris; Prononcé dans sa séance publique du 1 décembre 1784*. Parma: Bodoni, 1805.

Mulot, François-Valentin. "Journal intime de l'abbé Mulot." Ed. Maurice Tourneux. *Mémoires de la Société de l'histoire de Paris et de l'Ile-de-France* 29 (1902): 19–124.

Musée, autorisé par le gouvernement, sous la protection de Monsieur et de Madame. N.p., n.d.

Musée autorisé par le gouvernement, sous la protection de Monsieur et de Madame. Paris: L'Imprimerie de Monsieur, 1784.

Musée de Monsieur, et de Monseigneur Comte d'Artois. (Prospectus). Paris: L'Imprimerie de Monsieur, 1785.

Nicoles, Jean. *La Baguette divinatoire, ou Verge de Jacob*. Orig. edn, 1693. Ed. Paul Chacornac. Paris: La Diffusion Scientifique, 1959.

Nollet, Jean Antoine. *L'Art des expériences; ou, Avis aux amateurs de la physique*. 2nd edn. 3 vols. Paris: P.E.G. Durand, 1770.

——. *Cours de physique expérimentale*. Paris: P.G. Le Mercier, 1735.

——. *Essai sur l'électricité des corps*. Paris: Chez les frères Guerin, 1746.

——. *Leçons de physique expérimentale*. 6 vols. Amsterdam & Leipzig: Arkstee & Merkus, 1754–1765.

——. *Lectures in Experimental Philosophy*. Trans. John Colson. London: J. Wren, 1752.

——. *Lettres sur l'électricité*. 2 vols. Paris: H.-L. Guerin & L.-F. De la Tour, 1753–1760.

——. *Oratio habita a Joanne-Antonio Nollet, Licentiato Theologo, Regiae Scientiarum Academiae Socio, cùm primum Physicae Experimentalis Cursum Professor a Regie institutus auspicia retur*. Paris: Regis, 1753.

——. *Programme ou idée générale d'un cours de physique expérimentale avec un catalog raisonné des Instrumens qui servent aux expériences*. Paris: P.G. Le Mercier, 1738.

——, and Jean Jallabert. *Théories électriques du XVIIIe siècle: Correspondance entrel'abbé Nollet (1700–1770) et le physicien genevois Jean Jallabert (1712–1768)*. Ed. Isaac Benguigui. Geneva: Georg, 1984.

Nougaret, Pierre-Jean-Baptiste. *Tableau mouvant de Paris, ou, Variétés amusantes*. 3 vols. Paris: Duchesne, 1787.

Nouvelles de la république des lettres et des arts. Ed. Pahin de La Blancherie. Paris, 1779–1786.

d'Oberkirch, Henriette-Louise Waldner de Freundstein, baronne. *Mémoires de la Baronne d'Oberkirch sur la cour de Louis XVI et la sociéte française avant 1789*. Ed. Suzanne Burkard. Paris: Mercure de France, 1970.

Ordonnance de Police, qui fait défenses à toutes personnes de quelque qualité & condition qu'elles

soient, de fabriquer & faire enlever aucuns ballons. Angers: Mame, 1784.
Ordonnance de Police, qui fait défenses de fabriquer & faire enlever des Ballons. Paris: Pierres, 1784.
Ozanam, Jacques. *Récréations mathématiques et physiques*. 2 vols. Paris: Jean Jombert, 1694.
Paganiol de la Force, Jean-Aymar. *Description historique de la ville de Paris et de ses environs*. 10 vols. Paris: Chez les libraires associés, 1765.
Pahin de la Blancherie, Claude-Mammès. *Correspondance générale sur les sciences et les arts*. Paris: Au bureau de la Correspondance, 1779.
Panthot, Jean-Baptiste. *Lettre de M. Panthot*. Grenoble, 1692.
——. *Traité de la Baguette*, 3rd edn, Lyon: Thomas Amaulri & Jacques Guerrier, 1693.
Paulet, Jean Jacques. *Réponse à l'auteur des doutes d'un provincial, proposés à MM. les Médecins-Commissaires, chargés par le Roi de l'examen du magnétisme animal*. London, 1785.
Pelletier, François. *A MM. du District*. N.p., n.d.
——. *Exposé succinct des torts du sieur Alexandre Barré, ancien Garçon Boucher, aujourd'hui Capitaine des Grenadiers du District de l'Isle Saint-Louis, & Membre du Comité Militaire de l'Hôtel-de-Ville de Paris. Envers le Sieur Pelletier, Ingénieur-Machiniste Pensionné du Roi d'Espagne, avec un court abrégé de la vie du premier*. 1790.
——. *Hommage aux amateurs des arts*. Paris: Thiboust, 1782.
——. *Nouveau cabinet du sieur Pelletier*. Paris: Thiboust, 1784.
——. *Vente du cabinet du sieur Pelletier*. Paris: Herissant, 1785.
Perrin, M. *Amusemens de physique*. Paris: P. de Lormel, 1787a.
——. *Amusemens de physique*. Paris: P. de Lormel, 1787b.
——. *Amusemens physiques*. Paris: P. de Lormel, 1789.
——. *Nouveau spectacle*. Paris: P. de Lormel, 1785.
——. *Prospectus*. Paris: P. de Lormel, 1786.
——. *Prospectus*. Paris: P. de Lormel, 1787c.
Petit journal du Palais Royal, ou Affiches, annonces, et avis divers. Paris: Caveau, 1789.
Le Petit Tableau de Paris. Paris, 1783.
Philalète, *Lettre de l'hermite de Nivolet sur l'expérience aérostatique faite à Chambéry, le 22 Avril 1784*. N.p., 1784.
Pilâtre de Rozier, Jean François. *Première expérience de la Montgolfiére, construite par ordre du roi*. Paris: L'Imprimierie de Monsieur, 1784.
Pinetti, Jean-Joseph [Giuseppe]. *Amusemens physiques, et différentes expériences divertissantes*. 3rd edn. Paris: Gattey, 1791.
Pingeron, Jean-Claude. *L'Art de faire soi-même les ballons aérostatiques, conformes à ceux de M. de Montgolfier*. Paris: Hardouin, 1783.
Polinière, Pierre. *Expériences de physique*. Paris: J. de Laulne, 1709.
——. *Expériences de physique*, 5th edn, 2 vols. Paris: Clousier, 1741.
Poncelet, Polycarpe. *Principes généraux pour servir à l'éducation des enfans, particulièrement de la noblesse françoise*. 3 vols. Paris: P.G. Le Mercier, 1763.
Priestley, Joseph. *The History and Present State of Electricity*. 2 vols. Ed. Robert E. Schofield. 1775. 3rd edn. Reprint. New York: Johnson Reprint Corporation, 1966.

Procès-verbaux du Comité d'instruction publique de la Convention nationale. Ed. M.J. Guillaume. 6 vols. Paris: Imprimerie Nationale, 1891–1907.

Procès-verbaux et details des deux voyages aériens faits d'après la découverte de MM. Montgolfier. Bruxelles: Bailly, 1783.

Programme du Lycée, pour l'Année 1788. N.p., n.d.

Programme du Lycée, pour l'Année 1789. N.p., n.d.

"Programme du Lycée Républicain pour l'an V." *La Révolution française* 14 (1888): 1101–10.

Programme du Lycée Républicain. Pour la neuvième année Lycéenne. Paris: Perlot, n.d.

Pujoulx, Jean-Baptiste. *Paris à la fin du XVIIIe siècle*. Paris: Brigite Mathé, 1801.

Questions et conjectures sur l'application de l'électricité à l'aérostatique, aux aréostations, et à l'aérostation. Rodez: Marin Devic, 1786.

Rabaut-Saint-Etienne, Jean-Paul. "Lettre sur la vie et les écrits de M. Court de Gébelin, adressée au Musée de Paris." In *Oeuvres de Rabaut-Saint-Etienne*. 2 vols. Paris: Laisne, 1826: II:355–90.

Rabiqueau, Charles. *Comme l'électricité fait partie des semaines physiques & du cabinet de M. Rabiqueau*. Paris: Cailleau, 1772.

———. *Description de l'école de la vision, ou Cours sur le Livre du Microscope moderne*. Paris: Chez l'Auteur, 1783.

———. *Lettre contre l'électricité médicale*. Paris, 1772.

———. *Lettre électrique sur la mort de M. Richmann*. N.p., 1754.

———. *Lettre et regrets de souscription d'une jeune provinciale à une de ses amies à Paris*. Lyon, 1769.

———. *Manifeste littéraire, Servant de supplément aux Journaux sur le livre du Microscope moderne*. Paris: Chez l'Auteur, 1781.

———. *Le Microscope moderne*. Paris: Belin, 1785.

———. *Nouveau manège méchanique*. Paris: Chez l'Auteur, 1778.

———. *Prospectus du cabinet de M. Rabiqueau*. Paris: Chez l'Auteur, 1772.

———. *Relation curieuse et interessante pour le progrès de la physique et de la médecine*. Paris, 1760.

———. *Le Spectacle du feu élémentaire, ou cours d'électricité expérimentale*. Paris: Jombert, Knapen, & Duchesne, 1753.

[Rancy, de.] *Essai de physique en forme de lettres, à l'usage des jeunes personnes de l'un & l'autre sexes*. Paris: Hérissant, 1768.

[———.] *Lettres physiques, contenant les notions les plus nécessaires à ceux qui veulent suivre les Leçons expérimentales de cette science*. Paris: L.-G. de Hansy, 1763.

Règlements du Lycée Républicain. N.p., n.d.

Réimpression de l'ancien Moniteur: seule histoire authentique et inaltérée de la révolution française depuis la réunion des États-généraux jusqu'au Consulat (mai 1789–novembre 1799). Paris: Plon, 1858–1870.

Rivarol, Antoine de. *Lettre à Monsieur le Président de ***. Sur le Globe Airostatique, sur les Têtes parlantes, & sur l'état présent de l'opinion publique à Paris*. London, Paris: Cailleau, 1783.

Robert, Anne-Jean and Nicolas-Louis Robert. *Mémoire sur les expériences aérostatiques*.

Paris: L'Imprimerie de P.-D. Pierres, 1784.
Rohault, Jacques, *Traité de physique*, 2 vols. Paris, 1671.
——. *Traité de physique*. 2 vols. 6th edn. Paris: G. Desprez, 1683.
Rosnay, de. *La Physique des dames, ou les quatre élémens; ouvrage utile pour disposer à l'intelligence des Merveilles de la Nature*. Paris: Stoupe, 1773.
Rouland, M. *Déscription des machines électriques à taffetas, et leurs effets & des divers avantages que présentent ces nouveaux appareils*. Paris: Gueffier, 1785.
——. *Tableau historique des propriétés et des phénomènes de l'air*. Paris: Gueffier, 1784.
Rousseau, Jean-Jacques. *Emile, or On Education*. Trans. and Ed. Allan Bloom. New York: Basic Books, 1979.
——. *Julie, or, The New Heloise*. Trans. Philip Stewart and Jean Vaché. Hanover NH: University Press of New England, 1997.
——. "Lettres écrites de la montagne." In *Oeuvres complètes*. Eds. Bernard Gagnebin and Marcel Raymond. Paris: Gallimard, 1959–. III: 683–897.
Rousseau, Thomas. *Lettre à M. ***, sur les spectacles des boulevards*. Paris, 1781.
[Roze de Chantoiseau.] *Essai sur l'Almanach général d'indication d'adresse personnelle et domicile fixe, des six corps, arts et métiers*. Paris: La veuve Duchesne, 1769.
Saint-Maurice, Arnauld de. *L'Observateur volant et le triomphe héroïque de la navigation aérienne, et des vésicatoires amusants et célestes, poëme en quatre chants*. Paris: Cussac, 1784.
Sauri, Abbé. *L'Hydroscope, et le ventriloque*. Amsterdam; Paris: Valade, 1772.
Savérien, Alexandre. *Dictionnaire universel de mathématique et de physique*. 2 vols. Paris: J. Rollin and C.-A. Jombert, 1753.
——. *Histoire des philosophes modernes*. 8 vols. Paris: Bleuet and Guillaume, 1773.
Ségur, Louis Philippe de. *Memoirs and Recollections*. London: H. Colburn, 1825.
Sigaud de la Fond, Joseph Aignan. *Calendrier intéressant . . . ou Almanach Physico-Economique*. 4 vols. Paris: Boullon, 1772–1780.
——. *De l'électricité médicale*. Paris: Delaplace & Goujon, 1803.
——. *Description et usage d'un cabinet de physique expérimentale*. 2 vols. Paris: Gueffier, 1775.
——. *Dictionnaire de physique*. 5 vols. Paris: Rue et Hôtel Serpente, 1781–1782.
——. *Dictionnaire des merveilles de la nature*. 2 vols. Paris: Chardon, 1781.
——. *Elémens de physique théorique et expérimentale, pour servir de suite à la Description & usage d'un cabinet de physique expérimentale*. 4 vols. Paris: P.F. Gueffier, 1777.
——. *Essai sur différents especes d'air, qu'on désigne sous le nom d'air fixe, pour servir de suite & de supplément aux élémens de physique du même auteur*. Paris: P.F. Gueffier, 1779.
——. *Leçons de physique expérimentale*. 2 vols. Paris: Des Ventes de la Doué, 1767.
——. *Lettre sur l'électricité médicale*. Amsterdam, 1771.
——. *Physique générale et physique particulière*. 5 vols. Part of the *Bibliothèque Universelle des Dames*. Paris: Rue et Hôtel Serpente, 1788–1789.
——. *Précis historique et expérimentale des phénomènes électriques depuis l'origine de cette découverte jusqu'à ce jour*. Paris: Rue et Hôtel Serpente, 1781.
——. *Traité de l'électricité*. Paris: Des Ventes de la Doué, 1771.
Statuts et règlemens du premier musée, autorisé par le gouvernement, sous la protection de

Monsieur et de Madame, établi en 1781, par M. Pilatre de Rozier. N.p., n.d.

Supplément à l'art de voyager dans les airs, contenant le Précis historique de la grande Expérience faite à Lyon le 19 Janvier 1784 et l'Exposé d'un moyen ingénieux pour diriger à volonté les Ballons aérostatiques. N.p., n.d.

Teleki, Joseph. *La Cour de Louis XV: Journal de voyage du Comte Joseph Teleki*. Ed. Gabriel Tolnai. Paris: Presses Universitaires de France, 1943.

Tetu de Brissy, M. *Sur l'exposition de l'Expérience aérostatique faite le 18 juin 1786*. Paris, 1786.

Thiéry, Luc-Vincent. *Almanach du voyageur à Paris*. 5 vols. Paris: Hardouin, 1783–1787.

———. *Guide des amateurs et des étrangers voyageurs à Paris*. 2 vols. Paris: Hardouin & Gattey, 1786–1787.

Thouvenel, Pierre. *Mémoire physique et médicinal, montrant des rapports évidens entre les phénomenes de la baguette divinatoire, du magnétisme et de l'électricité*. 2 vols. Paris: Didot, 1781–1784.

Tournon de la Chapelle, Antoine. *La Vie et les mémoires de Pilâtre de Rozier, Ecrits par lui-même*. Paris: Belin, 1786.

Tressan, Louis Elisabeth de la Vergne de Broussin, comte de. *Souvenirs du comte de Tressan*. Versailles: Henry Lebon, 1897.

Vallemont, Pierre le Lorraine de. *La Physique occulte, ou Traité de la baguette divinatoire*. Amsterdam: Adrian Braakman, 1693.

———. *La Physique occulte, ou Traité de la baguette divinatoire . . . augumentée de plusieurs pièces*. Paris: Jean Coudot, 1696.

———. *La Physique occulte, ou, Traité de la baguette divinatoire*, 2 vols. The Hague: Moetgens, 1747.

Villard, Jean. *Lettre de M. Villard, ancien navigateur, mathématicien & ballo-metre, à Mge. de Flesselles, Intendant de la ville de Lyon*. Lyon: Imprimerie de Ville, 1784.

Viollet, Pierre. *Traité en forme de lettre contre la nouvelle rhabdomancie*. Lyons: H. Baritel, 1694.

Voltaire, François-Marie Arouet de. *Correspondance*. Ed. Theodore Besterman. 107 vols. Geneva: Institut et Musée Voltaire, 1965.

———. *Eléments de la philosophie de Newton*. Eds. Robert L. Walters and W.H. Barber. In *The Complete Works of Voltaire*. Eds. W.H. Barber and Ulla Kölving. Vol. 15. Oxford: The Alden Press, 1992.

———. *Letters on England*. Trans. Leonard Tancock. London: Penguin, 1980.

———. *Lettres Philosophiques*. Ed. F.A. Taylor. Oxford: Basil Blackwell, 1946.

———. *Notebooks*. Ed. Theodore Besterman. In *The Complete Works of Voltaire*. Eds. W.H. Barber and Ulla Kölving. Vols. 81/82. Oxford: Alden, 1968.

Secondary sources

Amiable, Louis. "Les Origines maçonniques du Musée de Paris et du Lycée." *La Révolution française* 31 (1896): 484–500.

Augarde, Jean-Dominique. "La Fabrication des instruments scientifiques du XVIIIe siècle et la corporation des fondeurs." In *Studies in the History of Scientific Instruments*. Eds. Christine Blondel, Françoise Parot, Anthony Turner, and Mari Williams. London: Rogers Turner, 1989: 52–72.

Baker, Keith Michael. *Condorcet: From Natural Philosophy to Social Mathematics*. Chicago: University of Chicago Press, 1975.

——. *Inventing the French Revolution*. Cambridge: Cambridge University Press, 1990.

Barrett, William F. "On the So-Called Divining Rod, or Virgula Divina." *Journal for Psychical Research* 13 (1898): 2–282 and 15 (1900): 130–383.

—— and Theodore Besterman. *The Divining Rod*. Toronto: Coles, 1979.

Barrière, Pierre. *L'Académie de Bordeaux: Centre de culture internationale au XVIIIe siècle (1712–1792)*. Bordeaux: Bière, 1951.

Becker, Carl L. *The Heavenly City of the Eighteenth-Century Philosophers*. New Haven: Yale University Press, 1932.

Benedict, Barbara M. *Curiosity: A Cultural History of Early Modern Inquiry*. Chicago: University of Chicago Press, 2001.

Benguigui, Isaac. "La Théorie de l'électricité de Nollet et son application en médecine à travers sa correspondance inédite avec Jallabert." *Gesnerus* 38 (1981): 225–35.

Benhamou, Reed. "Continuing Education and Other Innovations: An Eighteenth-Century Case Study." *Studies in Eighteenth-Century Culture* 15 (1986): 67–76.

——. "Cours publics: Elective Education in the Eighteenth Century." *Studies on Voltaire and the Eighteenth Century* 241 (1986): 365–76.

Bensaude-Vincent, Bernadette. *L'Opinion publique et la science: A chacun son ignorance*. Paris: Institut d'édition Sanofi-Synthélabo, 2000.

Bernard, Jean, Jean-François Lemaire, and Jean-Pierre Poirier, eds. *Marat homme de science?* Paris: Les Empêcheurs de Penser en Rond, 1993.

Besterman, Theodore. *Water-Divining: New Facts and Theories*. London: Methuen, 1938.

Bourdieu, Pierre. *Distinction: A Social Critique of the Judgement of Taste*. Trans. Richard Nice. Cambridge, MA: Harvard University Press, 1984.

Braunrot, Christabel P. and Kathleen Hardesty Doig. "The *Encyclopédie méthodique*: An Introduction." *Studies on Voltaire and the Eighteenth Century* 327 (1995): 1–152.

Bret, Patrice. "Un bateleur de la science: Le 'machiniste-physicien' François Bienvenu et la diffusion de Franklin et Lavoisier." *Annales historiques de la Révolution française* 338(4) (2004): 95–127.

Brewer, John and Roy Porter, eds. *Consumption and the World of Goods*. London: Routledge, 1993.

Brian, Eric. *La mesure de l'état: Administrateurs et géomètres au XVIIIe siècle*. Paris: Albin Michel, 1994.

Brockliss, Laurence W.B. "Aristotle, Descartes and the New Science: Natural Philosophy at the University of Paris, 1600–1740." *Annals of Science* 38 (1981): 33–69.

——. *French Higher Education in the Seventeenth and Eighteenth Centuries: A Cultural History*. Oxford: Clarendon, 1987.

—— and Colin Jones. *The Medical World of Early Modern France*. Oxford: Clarendon, 1997.

Broman, Thomas. "The Habermasian Public Sphere and Eighteenth-Century Historiography: A New Look at 'Science *in* the Enlightenment.'" *History of Science* 36 (1998): 123–49.

Brunet, Pierre. *L'Introduction des théories de Newton en France au XVIIIe siècle*. Geneva: Slatkine Reprints, 1970.

———. *Les physiciens hollandais et la méthode expérimentale en France au XVIIIe siècle*. Paris: Albert Blanchard, 1926.

Burke, Janet M. "Freemasonry, Friendship and Noblewomen: The Role of the Secret Society in Bringing Enlightenment Thought to Pre-Revolutionary Women Elites." *History of European Ideas* 10 (1989): 283–93.

———, and Margaret C. Jacob. "French Freemasonry, Women, and Feminist Scholarship." *Journal of Modern History* 68 (1996): 513–49.

Cabanes, Charles. "Histoire du premier musée autorisé par le gouvernement." *La Nature* (1937): 577–83.

———. "La Mort d'Icare: Pilâtre de Rozier." *La Nature* (1936): 529–33.

Cabrière, Justin. *Court de Gébelin, défenseur des églises réformées de France (1763–1784)*. Paris: Cahors, 1899.

Campardon, Emile. *Les Spectacles de la foire*. 2 vols. Paris: Berger-Levrault, 1877.

Cassirer, Ernst. *The Philosophy of the Enlightenment*. Trans. Fritz C.A. Koelln and James P. Pettegrove. Princeton: Princeton University Press, 1951.

Cazenove, Raoul de. *Premiers voyages aériens à Lyon en 1784*. Lyon: Pitrat, 1887.

Censer, Jack R. *The French Press in the Age of Enlightenment*. London: Routledge, 1994.

Chabaud, Gilles. "Entre sciences et sociabilités: Les Expériences de l'illusion artificielle en France à la fin du XVIIIe siècle." *Bulletin de la Société d'Histoire Moderne et Contemporaine* 44 (1997): 36–44.

———. "La Physique amusante et les jeux expérimentaux en France au XVIIIeme siècle." *Ludica* 2 (1996): 61–73.

———. "Sciences, magie et illusion: les romans de la physique amusante (1784–1789)." *Tapis-Franc* 8 (1997): 18–37.

Chartier, Roger. *Cultural History: Between Practices and Representations*. Trans. Lydia G. Cochrane. Ithaca: Cornell University Press, 1988.

———. *The Cultural Origins of the French Revolution*. Trans. Lydia G. Cochrane. Durham NC: Duke University Press, 1991.

———. "Popular Appropriations: The Readers and Their Books." In his *Forms and Meanings: Texts, Performances, and Audiences from Codex to Computer*. Philadelphia: University of Pennsylvania Press, 1995: 83–97.

Cherrière. "La lutte contre l'incendie dans les Halles, les marchés et les foires de Paris sous l'ancien régime." *Mémoires et documents pour servir à l'histoire du commerce et de l'industrie en France* 3 (1913): 107–321.

Chevreul, Michel-Eugène. *De la baguette divinatoire*. Paris: Mallet-Bachelier, 1854.

Chisick, Harvey. *The Limits of Reform in the Enlightenment*. Princeton: Princeton University Press, 1981.

Clair, Pierre. *Jacques Rohault (1618–1672): Bio-bibliographie*. Paris: Editions du Centre National de la Recherche Scientifique, 1978.

Clark, William, Jan Golinski, and Simon Schaffer, eds. *The Sciences in Enlightened Europe*. Chicago: University of Chicago Press, 1999.

Cohen, I. Bernard. "The Eighteenth-Century Origins of the Concept of the Scientific Revolution." *Journal of the History of Ideas* 37 (1976): 257–88.

——. *Franklin and Newton*. Philadelphia: The American Philosophical Society, 1956.

Cole, Arthur H. and George B. Watts. *The Handicrafts of France as Recorded in the Descriptions des Arts et Métiers, 1761–1788*. Kress Library of Business and Economics, vol. 8. Boston: Harvard Graduate School of Business Administration, 1952.

Coley, Awen A.M. "Followers of Daedalus: Science and Other Influences in the Tales of Flight in Eighteenth-Century French Literature." *Studies on Voltaire and the Eighteenth Century* 371 (1999): 81–173.

Conner, Clifford Despard. *Jean Paul Marat: Scientist and Revolutionary*. New Jersey: Humanities Press, 1997.

Cooter, Roger and Stephen Pumfrey. "Separate Spheres and Public Places: Reflections on the History of Science Popularization and Science in Popular Culture." *History of Science* 32 (1994): 237–67.

Corson, David W. "Pierre Polinière, Francis Hauksbee, and Electroluminescence: A Case of Simultaneous Discovery." *Isis* 59 (1968): 402–13.

Costabel, Pierre and Monette Martinet. *Quelques savants et amateurs de Science au XVIIe siècle*. Paris: Société Française d'Histoire des Sciences et des Techniques, 1986.

Coutil, Léon. *Jean-Pierre Blanchard, physicien-aéronaute*. Evreux: Charles Herissey, 1911.

Crosland, Maurice. "The Development of a Professional Career in Science in France." *Minerva* 13 (1975): 38–57.

Darnton, Robert. *The Literary Underground of the Old Regime*. Cambridge, MA: Harvard University Press, 1982.

——. *Mesmerism and the End of the Enlightenment in France*. Cambridge, MA: Harvard University Press, 1968.

Daumas, Maurice. *Les cabinets de physique au XVIIIe siècle*. Paris: Conférence du Palais de la Découverte, 1951.

——. *Scientific Instruments of the Seventeenth and Eighteenth Centuries and Their Makers*. Trans. Mary Holbrook. London: Portman Books, 1972.

Dejob, Charles. "De l'éstablissement connu sous le nom de Lycée et d'Athénée et de quelques établissements analogues." *Revue internationale de l'enseignement* 18 (1889): 4–38.

Delon, Michel. "La marquise et la philosophe." *Revue des sciences humaines* 182 (1981): 65–78.

Dhombres, Nicole and Jean Dhombres. *Naissance d'un pouvoir: sciences et savants en France (1793–1824)*. Paris: Payot, 1989.

Dorveaux, Paul. "Pilâtre de Rozier." *Bulletin de la Société d'Histoire de la Pharmacie* (1920): 249–58.

——. "Pilâtre de Rozier et l'Académie des Sciences." *Les Cahiers lorrains* 8 (1929): 162–6, 182–5.

Douglas, Aileen. "Popular Science and the Representation of Women: Fontenelle and After." *Eighteenth-Century Life* 18 (1994): 1–14.

Douthwaite, Julia. *The Wild Girl, Natural Man, and the Monster: Dangerous Experiments in the Age of Enlightenment*. Chicago: University of Chicago Press, 2002.

Duval, Clément. "Pilâtre de Rozier (1754–1785): Chemist and First Aeronaut." *Chymia* 12 (1967): 99–117.

Fara, Patricia. *Sympathetic Attractions: Magnetic Practices, Beliefs, and Symbolism in Eighteenth-Century England*. Princeton: Princeton University Press, 1996.

Fehér, Marta. "The Triumphal March of a Paradigm: A Case Study of the Popularization of Newtonian Science." *Tractrix* 2 (1990): 93–110.

Feyel, Gilles. "Médecins, empiriques et charlatans dans la presse provinciale à la fin du XVIIIe siècle." In *Le Corps et la santé: Actes du 110e congrès national des sociétés savantes*. Paris: C.T.H.S., 1985: 79–100.

Figuier, Louis. *Histoire du merveilleux dans les temps modernes*, 4 vols. Paris: Hachette, 1860.

Findlen, Paula. "Translating the New Science: Women and the Circulation of Knowledge in Enlightenment Italy." *Configurations* 3 (1995): 167–206.

France, Anatole. *L'Elvire de Lamartine: Notes sur M. & Mme. Charles*. Paris: H. Champion, 1893.

Freudenthal, Gad. "Early Electricity Between Chemistry and Physics: The Simultaneous Itineraries f Francis Hauksbee, Samuel Weil, and Pierre Polinière," *Historical Studies in the Physical Sciences* 11 (1981): 203–29.

———. "Littérature et sciences de la nature en France au début du XVIIIe siècle." *Revue de synthèse* 100 (1980): 267–95.

Garrioch, David. *The Making of Revolutionary Paris*. Berkeley: University of California Press, 2002: 265.

———. *The Formation of the Parisian Bourgeoisie, 1690–1830*. Cambridge, MA: Harvard University Press, 1996.

Gilardin, Alphonse. "Un procès à Lyon en 1692, ou Aymar, l'homme à la baguette." *Revue du Lyonnais* (1837): 81–99.

Gillespie, Richard. "Ballooning in France and Britain, 1783–1786: Aerostation and Adventurism," *Isis* 75 (1984): 249–68.

Gillispie, Charles C. *The Montgolfier Brothers and the Invention of Aviation: 1783–1784*. Princeton: Princeton University Press, 1983.

———. *Science and Polity in France at the End of the Old Regime*. Princeton: Princeton University Press, 1980.

———. *Science and Polity in France: The Revolutionary and Napoleonic Years*. Princeton: Princeton University Press, 2004.

Goldgar, Anne. *Impolite Learning: Conduct and Community in the Republic of Letters, 1680–1750*. New Haven: Yale University Press, 1995.

Golinski, Jan. *Science as Public Culture: Chemistry and Enlightenment in Britain, 1760–1820*. Cambridge: Cambridge University Press, 1992.

Goodman, Dena. "Public Sphere and Private Life: Toward a Synthesis of Current Historiographical Approaches to the Old Regime." *History and Theory* 31 (1992):

1–20.

———. *The Republic of Letters: A Cultural History of the French Enlightenment*. Ithaca: Cornell University Press, 1994.

Guénot, Hervé. “La Correspondance générale pour les Sciences et les Arts de Pahin de La Blancherie (1779–1788).” *Cahiers Haut-Marnais* 162 (1985): 49–61.

———. “Musées et lycées parisiens (1780–1830).” *Dix-huitième siècle* 18 (1986): 249–67.

———. “Une nouvelle sociabilité savante: Le Lycée des Arts.” In *La Carmagnole des muses: L'homme de lettres et l'artiste dans la Révolution*. Paris: Armand Colin, 1988: 67–78.

Habermas, Jürgen. *The Structural Transformation of the Bourgeois Public Sphere: An Inquiry into a Category of Bourgeois Society*. Trans. Thomas Burger. Cambridge, MA: MIT Press, 1989.

Hahn, Roger. *The Anatomy of a Scientific Institution: The Paris Academy of Sciences, 1666–1803*. Berkeley: University of California Press, 1971.

———. “Changing Patterns of Support of Scientists from Louis XIV to Napoleon.” *History and Technology* 4 (1987): 401–11.

Haines, Barbara. “The Athénée de Paris and the Bourbon Restoration.” *History and Technology* 5 (1988): 249–71.

Hankins, Thomas. *Science and the Enlightenment*. Cambridge: Cambridge University Press, 1985.

Hanna, Blake T. “Polinière and the Teaching of Experimental Physics in Paris: 1700–1730.” In *Eighteenth-Century Studies Presented to Arthur M. Wilson*. Ed. Peter Gay. Hanover NH: University Press of New England, 1972: 13–39.

Hazard, Paul. *The European Mind: The Critical Years, 1680–1715*. Trans. J. Lewis May. New York: Fordham University Press, 1990.

Heilbron, John L. *Electricity in the Seventeenth and Eighteenth Centuries: A Study of Early Modern Physics*. Berkeley: University of California Press, 1979.

Hillaire-Pérez, Lilliane. *L'Invention technique au siècle des Lumières*. Paris: Albin Michel, 2000.

Hochadel, Oliver. *Öffentliche Wissenschaft: Elektrizität in der deutschen Aufklärung*. Göttingen: Wallstein, 2003.

Home, Roderick Weir. “Electricity in France in the Post-Franklin Era.” In *Proceedings of The XIVth International Congress of the History of Science*. Tokyo: Science Council of Japan, 1975: 1–4.

———. “The Notion of Experimental Physics in Early Eighteenth-Century France.” In *Change and Progress in Modern Science*. Ed. Joseph C. Pitt. Dordrecht: D. Reidel, 1985: 107–31.

———. “Out of a Newtonian Straitjacket: Alternative Approaches to Eighteenth-Century Physical Science.” *Studies in the Eighteenth Century* 4 (1979): 235–49.

Hunn, James Martin. “The Balloon Craze in France, 1783–1799: A Study in Popular Science.” Ph.D. diss., Vanderbilt University, 1982.

Isherwood, Robert M. *Farce and Fantasy: Popular Entertainment in Eighteenth-Century Paris*. Oxford: Oxford University Press, 1986.

Jacob, Margaret. *Living the Enlightenment: Freemasonry and Politics in Eighteenth-Century*

Europe. Oxford: Oxford University Press, 1991.

Jones, Colin. "The Great Chain of Buying: Medical Advertisement, the Bourgeois Public Sphere and the Origins of the French Revolution." *American Historical Review* 101 (1996): 13–40.

——. "Pulling Teeth in Eighteenth-Century Paris." *Past and Present* 166 (2000): 100–46.

Kaplow, Jeffrey. *The Names of Kings: The Parisian Laboring Poor in the Eighteenth Century*. New York: Basic Books, 1972.

Kirsop, Wallace. "Cultural Networks in Pre-Revolutionary France: Some Reflexions on the Case of Antoine Court de Gébelin." *Australian Journal of French Studies* 18 (1981): 231–47.

Kleinert, Andreas. *Die Allgemeinverständlichen Physikbücher der französischen Aufklärung*. Aarau: Sauerländer, 1974.

——. "La vulgarisation de la physique au siècle des lumières." *Francia* 10 (1982): 303–12.

Koerner, Lisbet. "Women and Utility in Enlightenment Science." *Configurations* 3 (1995): 233–55.

Koselleck, Reinhart. *Critique and Crisis: Enlightenment and the Pathogenesis of Modern Society*. Cambridge, MA: MIT Press, 1988.

Lecot, Victor-Lucien-Sulpice. *L'abbé Nollet de Pimprez*. Noyon: Cottu-Harlay, 1856.

Licoppe, Christian. *La Formation de la pratique scientifique: le discours de l'expérience en France et en Angleterre (1630–1820)*. Paris: Editions la découverte, 1996.

Lough, John. *Paris Theatre Audiences in the Seventeenth & Eighteenth Centuries*. London: Oxford University Press, 1957.

Mason, Haydon Trevor. "Algarotti and Voltaire." In *Mélanges à la mémoire de Franco Simone*. 3 vols. Geneva: Slatkine, 1981: II: 467–80.

Maza, Sarah. *Private Lives and Public Affairs: The Causes Célèbres of Prerevolutionary France*. Berkeley: University of California Press, 1993.

——. *The Myth of the French Bourgeoisie: An Essay on the Social Imaginary, 1750–1850*. Cambridge, MA: Harvard University Press, 2003.

McClaughlin, Trevor. "Censorship and Defenders of the Cartesian Faith in Mid-Seventeenth Century France." *Journal of the History of Ideas* 40 (1979): 563–81.

——. "Le Concept de science chez Jacques Rohault." *Revue d'histoire des sciences* 30 (1977): 225–40.

McClellan, James E. *Colonialism and Science: Saint Domingue in the Old Regime*. Baltimore: Johns Hopkins University Press, 1992.

——. *Science Reorganized: Scientific Societies in the Eighteenth Century*. New York: Columbia University Press, 1985.

McCracken, Grant. *Culture and Consumption: New Approaches to the Symbolic Character of Consumer Goods and Activities*. Bloomington: Indiana University Press, 1988.

McKendrick, Neil, John Brewer, and J.H. Plumb. *The Birth of a Consumer Society: The Commercialization of Eighteenth-Century England*. Bloomington: Indiana University Press, 1985.

Melton, James Van Horn. *The Rise of the Public in Enlightenment Europe*. Cambridge:

Cambridge University Press, 2001.

Morman, Paul J. "Rationalism and the Occult: The 1692 Case of Jacques Aymar, Dowser *Par Excellence*." *Journal of Popular Culture* 19 (1986): 119–29.

Mornet, Daniel. *Les origines intellectuelles de la Révolution Française, 1715–1787*. Paris: Armand Colin, 1967 [1933].

——. *Les Sciences de la nature en France, au XVIIIe siècle*. Paris, 1911.

Mouy, Paul. *Le Développement de la physique cartésienne, 1646–1712*. Paris: Vrin, 1934.

Pacaut, M. "Le physicien, Jacques Rohault (1620–1672)." *Memoires de l'Académie des sciences, des lettres et des arts d'Amiens* 8 (1881): 1–26.

Peiffer, Jeanne. "L'Engouement des femmes pour les sciences au XVIIIe siècle." In *Femmes et pouvoirs sous l'ancien régime*, eds. Danielle Haase-Dubosc and Eliane Viennot. Paris: Editions Rivages, 1991: 196–222.

——. La Littérature scientifique pour les femmes au siècle des lumières." In *Sexe et genre: De la hiérarchie entre les sexes*, eds. Marie-Claude Hurtig, Michèle Kail, and Hélène Rouch. Paris: Editions du Centre National de la Recherche Scientifique, 1991: 137–46.

Porter, Roy, ed., *The Cambridge History of Science*. Vol. 4. *Eighteenth-Century Science*. Cambridge: Cambridge University Press, 2003.

——. "Science, Provincial Culture and Public Opinion in Enlightenment England." *British Journal for Eighteenth Century Studies* 3 (1980): 20–46.

Poterlet (jeune), M. *Notice sur Madame Blanchard, aéronaute*. Paris: Imprimerie de Fain, 1819.

Quignon, G. Hector. "L'abbé Nollet, physicien: son voyage en Piémont et en Italie (1749)." *Mémoires de l'Académie d'Amiens* 51 (1904): 473–539.

Rabbe, Félix. "Pahin de la Blancherie et le salon de la correspondance." *Bulletin de la Société Historique du VIe Arrondissement de Paris* 2 (1899): 30–52.

Raichvarg, Daniel and Jean Jacques. *Savants et ignorants: une histoire de la vulgarisation des sciences*. Paris: Editions du Seuil, 1991.

Ramsey, Matthew. *Professional and Popular Medicine in France, 1770–1830: The Social World of Medical Practice*. Cambridge: Cambridge University Press, 1988.

Ravel, Jeffrey. *The Contested Parterre: Public Theatre and French Political Culture, 1680–1791*. Ithaca: Cornell University Press, 1999.

Riskin, Jessica. *Science in the Age of Sensibility: The Sentimental Empiricists of the French Enlightenment*. Chicago: University of Chicago Press, 2002.

Robbins, Louise E. *Elephant Slaves and Pampered Parrots: Exotic Animals in Eighteenth-Century Paris*. Baltimore: Johns Hopkins University Press, 2002.

Robineau, Lucien. "Lazare Carnot et les compagnies d'aérostiers." *Revue historique des Armées* 2 (1989): 101–10.

Roche, Daniel ed. *Almanach parisien en faveur des étrangers et des personnes curieuses*. Saint-Etienne: Publications de l'Université de Saint-Etienne, 2001.

——. *The Culture of Clothing: Dress and Fashion in the 'Ancien Régime.'* Trans. Jean Birrell. Cambridge: Cambridge University Press, 1994.

——. *La France des lumières*. Paris: Fayard, 1993.

——. *A History of Everyday Things: The Birth of Consumption in France, 1600–1800*.

Trans. Brian Pearce. Cambridge: Cambridge University Press, 2000.
——. *The People of Paris: An Essay in Popular Culture in the Eighteenth Century*. Trans. Marie Evans with Gwynne Lewis. Berkeley: University of California Press, 1987.
——. *Le Siècle des lumières en province: académies et académiciens provinciaux, 1680–1789*. 2 vols. Paris: Mouton, 1978.
Root-Bernstein, Michele. *Boulevard Theater and Revolution in Eighteenth-Century Paris*. Ann Arbor, MI: UMI Research Press, 1984.
Rousseau, G.S. and Roy Porter, eds. *The Ferment of Knowledge: Studies in the Historiography of Eighteenth-Century Science*. Cambridge: Cambridge University Press, 1980.
Ruffet, Monique. "La physique pour débutants: Angélique Diderot et les leçons de l'abbé Nollet." *Recherches sur Diderot et sur l'Encyclopédie* 13 (1992): 57–78.
Rupp, Jan C.C. "The New Science and the Public Sphere in the Premodern Era." *Science in Context* 8 (1995): 487–507.
Sarton, George. "The Study of Early Scientific Textbooks." *Isis* 38 (1947): 137–48.
Schaffer, Simon. "Natural Philosophy and Public Spectacle in the Eighteenth Century." *History of Science* 21 (1983): 1–43.
Schiebinger, Londa. *The Mind Has No Sex? Women in the Origins of Modern Science*. Cambridge, MA: Harvard University Press, 1989.
Schneider, Rachel R. "Star Balloonist of Europe: The Career of Marie-Madeleine Blanchard." *Consortium on Revolutionary Europe, 1750–1850: Proceedings* (1983): 697–711.
Schwartz, Vanessa R. *Spectacular Realities: Early Mass Culture in Fin-de-Siècle Paris*. Berkeley: University of California Press, 1998.
Sgard, Jean, ed. *Dictionnaire des journaux, 1600–1789*. 2 vols. Paris: Universitas, 1991.
——. "Les philosophes en montgolfière." *Studies on Voltaire and the Eighteenth Century* 303 (1992): 99–111.
——, ed. *La Presse provinciale au XVIIIe siècle*. Grenoble: Centre de Recherches sur les Sensibilités, 1983.
Shank, J.B. "Before Voltaire: Newtonianism and the Origins of the Enlightenment in France." Ph.D. diss., Stanford University, 2000.
Shapin, Steven. "The Audience for Science in Eighteenth-Century Edinburgh." *History of Science* 12 (1974): 95–121.
Smeaton, William A. "The Early Years of the Lycée and the Lycée des Arts." *Annals of Science* 11 (1955): 257–67, 309–19.
——. "The First and Last Balloon Ascents of Pilâtre de Rozier." *Archives internationales d'histoire des sciences* 44 (1958): 263–69.
——. "Jean-François Pilâtre de Rozier, the First Aeronaut." *Annals of Science* 11 (1955): 349–55.
Smith, Woodruff D. *Consumption and the Making of Respectability, 1600–1800*. London:Routledge, 2002.
Sonenscher, Michael. *Work and Wages: Natural Law, Politics, and the Eighteenth-Century French Trades*. Cambridge: Cambridge University Press, 1989.

Sonnet, Martine. *L'éducation des filles au temps des lumières*. Paris: Les Editions du Cerf, 1987.

Spary, Emma. *Utopia's Garden: French Natural History from Old Regime to Revolution*. Chicago: University of Chicago Press, 2000.

Stafford, Barbara Maria. *Artful Science: Enlightenment Entertainment and the Eclipse of Visual Education*. Cambridge, MA: MIT Press, 1994.

Staum, Martin. "Physiognomy and Phrenology at the Paris Athénée." *Journal of the History of Ideas* 56 (1995): 443–62.

Stewart, Larry. "Public Lectures and Private Patronage in Newtonian England." *Isis* 77 (1986): 47–58.

——. *The Rise of Public Science: Rhetoric, Technology, and Natural Philosophy in Newtonian Britain, 1660–1750*. Cambridge: Cambridge University Press, 1992.

——. "The Selling of Newton: Science and Technology in Early Eighteenth-Century England." *Journal of British Studies* 25 (1986): 178–92.

Sturdy, David J. *Science and Social Status: The Members of the Académie des Sciences, 1666–1750*. New York: Boydell, 1995.

Sutton, Geoffrey V. "Electric Medicine and Mesmerism." *Isis* 72 (1981): 375–92.

——. *Science for a Polite Society: Gender, Culture, and the Demonstration of Enlightenment*. Boulder, CO: Westview Press, 1995.

Taton, René. "Madame du Châtelet, traductrice de Newton." *Archives internationales d'histoire des sciences* 89 (1969): 185–210.

——, ed. *Enseignement et diffusion des sciences en France au XVIIIe siècle*. Paris: Hermann, 1964.

Terrall, Mary. "Gendered Spaces, Gendered Audiences: Inside and Outside the Paris Academy of Sciences." *Configurations* 3 (1995): 207–32.

——. *The Man Who Flattened the Earth: Maupertuis and the Sciences in the Enlightenment*. Chicago: University of Chicago Press, 2002.

Thébaud-Sorger, Marie. "'L'air du temps.' L'aérostation: savoirs et pratiques à la fin du XVIIIe siècle (1783–1785)." Thèse du doctorat (Ecole des hautes études en sciences sociales), 2004.

Tissier, André. *Les spectacles à Paris pendant la Révolution*. Geneva: Droz, 1992.

Todd, Christopher. "French Advertising in the Eighteenth Century." *Studies on Voltaire and the Eighteenth Century* 266 (1989): 513–47.

Torlais, Jean. *L'abbé Nollet (1700–1770) et la physique expérimentale au XVIIIe siècle*. Paris: Conférences du Palais de la Découverte, 1958.

——. *Un physicien au siècle des lumières: L'abbé Nollet, 1700–1770*. Paris: Sipuco, 1954.

——. "Un prestidigitateur célèbre, chef de service d'électrothérapie au XVIIIe siècle, Ledru dit Comus (1731–1807)." *Histoire de la médecine* 5 (1955): 13–25.

Wade, Ira O. *The Intellectual Development of Voltaire*. Princeton: Princeton University Press, 1969.

Walters, Alice Nell. "Conversation Pieces: Science and Politeness in Eighteenth-Century England." *History of Science* 35 (1997): 121–54.

Wellman, Kathleen. *Making Science Social: The Conferences of Théophraste Renaudot, 1633–1642*. Norman: University of Oklahoma Press, 2003.

Wilkins, Kay S. "The Treatment of the Supernatural in the *Encyclopédie*." *Studies on Voltaire and the Eighteenth Century* 90 (1972): 1757–71.

Index